Vickers: Against the Odds
1956–1977

Vickers:
Against the Odds

1956–1977

Harold Evans

HODDER AND STOUGHTON
LONDON SYDNEY AUCKLAND TORONTO

British Library Cataloguing in Publication Data
Evans, Harold
    Vickers: Against the Odds
    1. Vickers Limited – History 2. Munitions – Great Britain – History
    1 Title
    388.7'62'340941   HD9743.G64V/

    ISBN 0–340–23434–2

Hodder and Stoughton Editorial Office: 47 Bedford Square, London WC1B 3DF.

Acknowledgments

THOUGH THE APPROACH OF A COMMERCIAL COMPANY TO THE KEEPING of records (other than financial records) is likely to be strongly different from that of the Civil Service—less methodical, less detailed, less consistent—the archives of Vickers Limited provide a rich source of documentation. Gaps certainly exist, and the thinking behind decisions is often not displayed as it would be in Civil Service documentation, but I have had no difficulty in finding source material about events and developments in the Company's affairs during the period of which I write. In seeking it I have been greatly helped by the Company Secretary, Hugh Scrope, who believes in maintaining records which are genuinely informative, and who knows from personal experience in writing about Vickers aircraft the problems and requirements of the researcher. From him, let it be added, came the proposal that this book should be written.

A principal source of information must obviously be the Company's annual reports and accounts, and Vickers had progressively enlarged the scope of its reports even before political pressures began to emerge to require this. In conjunction with each annual report goes the Chairman's address at the annual general meeting, reinforcing and bringing up to date his statement in the report.

There is a limit to the amount of detail that can be included in an annual report, however, and in delving more deeply I found particularly helpful the internal report on the accounts prepared annually for much of the period by the Chief Accountant, Malcolm Spriggs. Meetings of the Board, and of the executive committee, are generally full and informative, and the formal Board minutes were supported for some years by informal minutes which set out in rather more detail the pros and cons raised in discussion. On issues of particular moment memoranda were usually put before the Board to expose fully the considerations to be weighed. In relation to the 'intervention' by certain institutional stockholders in 1969, a detailed record was kept by Sir Leslie Rowan.

In addition to records of meetings of the Board of Vickers Limited, minutes of the boards of subsidiary groups and units supply information about the problems and decisions of the individual businesses. The Group Management Conferences are also well recorded. Head Office notices, announcing developments in organisation and senior appointments, are another useful source, as are *Vickers Magazine* and *Vickers News*. The magazine was discontinued in 1970 for reasons of economy, but over many years it had brought together articles in depth, some highly technical, on topics associated with Vickers activities. *Vickers News*, as its title indicates, is designed to provide a regular flow of company news to everyone working in the Company, though it also has some external circulation.

As a company for so long at the centre of British armaments design and manufacture, as also of the aircraft and shipbuilding industries, Vickers figures in Hansard and in Parliamentary and other government reports. Where such references are mentioned in the text the origin is stated.

Documentation, however full, does not necessarily give 'the feel' of events, and I am grateful to those involved in them who have been ready to talk to me at length and to draw on memory and on personal impressions in order that flesh may be put on the bones. I had particular help from R. P. H. Yapp, especially in sorting out the complexities involved in the formation of British Aircraft Corporation and consequential developments.

In writing about nuclear submarines, I was greatly helped, from outside the Vickers perimeter, by Sir Rowland Baker, who was a key figure in the official defence hierarchy at the critical times. The American development, first, of nuclear-propelled submarines, and then of submarines both nuclear-propelled and nuclear-armed, is well told in *The Atomic Submarine*, a biography of Admiral Hyman G. Rickover by Clay Blair Jr, and in *Adventure in Partnership: The Story of Polaris*, a publication sponsored by 'a number of Polaris industrial contractors who felt that a record of this kind should be compiled and distributed'. Another readable American account of submarine history is *The Far and the Deep* by Commander E. P. Stafford, USN. A useful British publication is Commander Nicholas Whitestone's *The Submarine: The Ultimate Weapon*. Vickers Shipbuilding Group customarily published excellent booklets to commemorate the commissioning of nuclear submarines (and other naval vessels), including *Polaris and the Royal Navy 1963–1973*, designed for the Ministry of Defence (Navy). Useful information also emerged from an exhibition at the National Maritime Museum, Greenwich, to mark *75 Years of British Submarines*

1901–1976. In referring to the Polaris agreement reached in principle during the meeting between President Kennedy and Mr Harold Macmillan at Nassau in December 1962, I was able to draw on personal experience as a member of the British delegation.

The chapter 'Down to the Sea Bed' tells in some detail of the rescue of the *Pisces* submersible off the southern Irish coast in 1973. The principal source of information about the accident and the rescue is the report of the official inquiry subsequently held under Vickers auspices *Pisces III Accident Inquiry*. An excellent personal account of the incident is given by Roger Chapman, one of the two men crewing *Pisces III*, in his book *No Time on Our Side*.

Though my task has been to write about Vickers during the period 1956–1977, I could obviously not do so effectively without reference to earlier developments. The prime source for these references is *Vickers: A History* by J. D. Scott, published in 1962. Two sources which Scott mentions, and I also acknowledge, are Hugh Scrope's *Golden Wings: The Story of Fifty Years of Aviation by the Vickers Group of Companies 1908–1958*, and another privately circulated Vickers publication *Vickers Limited Group of Companies: A Short Account*, compiled by V. F. G. Pritchett and completed in November 1959. For those interested in the movement by Vickers into armaments manufacture towards the end of the last century and the beginning of this, *The Vickers Brothers* by Clive Trebilcock, published in 1977, is a scholarly study. Interesting sidelights on this period are provided by Sir Hiram S. Maxim in his autobiography *My Life*, and by David Dougan in *The Great Gun-Maker*, a biography of Lord Armstrong. For an account of ships built at Barrow a book to see is *A Century of Shipbuilding* by Tom Clark, published in 1971. The fascinating history of the Cockatoo Dockyard at Sydney is told by Roger Parker in *Cockatoo Island*, published in 1977. The early days of the Roneo company are recorded in *The Roneo Story* by John Dorlay, who retired in 1970, when he was marketing director, after 39 years service with the company. The beginnings of W. H. Howson Limited were remembered for me by Gilbert Matthewman, one of Hector Howson's associates in starting the business and, in 1977, still serving with the company as special projects manager in the research department.

Finally, I place on record my warm appreciation of all the help I have received from Miss Ivy Law, my secretary for most of the time I was at Vickers House. To be added to high skills as a secretary was her close knowledge of the Company's organisation, and of Vickers House personalities, acquired particularly during her spells of duty as secretary to Lord Weeks and to Lord Knollys.

Contents

Illustrations

Most of the photographs are the work of Vickers Limited photographers, in particular Ian Stokes, the present chief photographer. The jacket illustration is by him and his other pictures include *Oriana*, *Pisces III*, the Canadian Vickers nuclear reactor and Vickers in London. Also represented is the work of the late Leslie Sansom (notably the helicopter view of Millbank Tower nearing completion), Brian Wexham, who succeeded him as head of the department, and Bob Fisher. Vickers Limited set up a Film and Photographic Department in 1916 and its work has won national and international recognition: it now operates as the Photographic and Audio-Visual Department.

Foreword

WHEN, IN 1963, VICKERS OBTAINED A GRANT OF ARMORIAL BEARINGS, the Directors were asked to approve a motto. They chose 'Service in Peace and War'. In many ears it must have sounded priggish (as mottoes so often do). In fact, it was a plain statement of attitude developed during nearly a century and a half of making steel, building ships and aircraft, and making arms for the Navy, Army and Air Force. Not the right attitude in a private enterprise company, it could be argued, and indeed was so argued by a group of the Company's institutional shareholders in 1969. The Vickers Board, remarked the City Editor of the *Daily Mail*, 'seemed more devoted to the national interest than to the shareholders'. There should be more 'sons of bitches' on the Board said the institutional group's spokesman. This might have been considered a little hard on a business which had never failed to make a profit in 100 years as a public company and had rarely paid a dividend less than 10 per cent: but, the argument went, if the management paid more attention to the interests of the stock-holders and were quicker to take tough decisions, then profits could be considerably bigger, dividends considerably higher and prospects considerably better.

Nor did a long record of successful 'service in peace and war' do anything to moderate the missionary zeal of post-war Labour Governments in the cause of State ownership and the direction of industry from Whitehall. Vickers had been built on four pillars – steel, engineering, shipbuilding and aircraft. Steel was swallowed (second bite) in 1967. Aircraft (Vimy, Spitfire, Wellington, Viking, Valiant, Viscount, VC10) became first a holding in the new British Aircraft Corporation in 1960 and then disappeared into British Aerospace in 1977. Shipbuilding ('The Royal Navy's senior shipbuilder', said the Queen in 1971) also went into State ownership in 1977.

With three of its four main pillars removed the Vickers edifice might have been expected to collapse. In fact, there had been time to build new pillars. The engineering pillar had been powerfully reinforced, not only in the United Kingdom, but also in Australia and Canada. A new engineering pillar—offshore engineering—had been introduced. Office equipment (Roneo Vickers, with a turnover exceeding £100m) constituted a third, and lithographic plates (Howson-Algraphy, a high profit maker) a fourth. By the second half of 1977 the edifice looked very different, and perhaps a new motto would be needed over the door, but its strength and stability were undeniable.

These events covered a period of rather more than 20 years. In writing about them I have been given unrestricted access to Vickers' records, and have talked to most of those principally involved. The assessments and interpretations are my own. They are probably coloured by a degree of personal participation during the last ten years. Nevertheless, I have sought to compile an account as objective as I could make it and one which I hope will prove valuable to anyone who writes about Vickers at a later date, with the advantages of time and perspective.

To cover such a diversity of activities and events over a long period poses obvious difficulties of presentation if, hopefully, a readable narrative is to result. I have tried to meet the problem by dividing the time span into four parts, and within each including a narrative chapter or chapters, supported by chapters on topics meriting more detailed treatment. This inevitably means a degree of repetition, but against this disadvantage I feel there is some advantage in having self-contained accounts of specific subjects.

There is one further point to be made. The major decisions in managing an industrial enterprise as large and diverse as Vickers must be taken at the centre and by a relatively few individuals. The characters and personalities of these individuals, and the interplay between them, are therefore of paramount importance and cannot be ignored. Necessarily I have had to try to give some outline to these personalities and once again the assessments and interpretations are my own, based on collated views, on the evidence of events, and in most instances on personal acquaintance.

In concentrating on events at the centre I could obviously not deal in detail with the affairs of particular groups and units. Many of these are sufficiently large and important to merit books in themselves, and indeed several such books already exist (and have provided me with valuable information) or are in preparation. For a full understanding of Vickers these works need to be consulted as well as the account I now present.

ROTTINGDEAN,
May, 1978.

Part One
1956

'A thriving and well-knit organisation'

The Times

'A Thriving and Well-knit Organisation'

EDWARD GEORGE WILLIAM TYRWHITT KNOLLYS, SECOND VISCOUNT, GCMG, MBE, DFC, 'Edgey' to his friends ('don't leave out the "e" will you?'), became Chairman of Vickers Limited in 1956.

Against his family and social background it might have seemed an odd appointment. His father, the first Viscount, had been Private Secretary to King Edward VII for 40 years and to King George V from 1910 to 1913. His godfather was George V. He himself had been a Royal Page. For several centuries before that the Knollys family had been upper crust. In 1381 'Sir Robert Knolles and Constance, his wife' were granted leave to build a Hautpas in Seething Lane in the City of London. The fact was that Lady Constance had already built it – Sir Robert being away at the wars – but no penalty was imposed beyond the payment annually on Midsummer Day of one red rose from her garden. Nearly six centuries later the Knollys rose was still being presented to the Lord Mayor on Midsummer Day.

The young 20th-century Knollys, however, compelled perhaps by the harsh necessities of the 1920s, saw no reason why someone born into the aristocracy should be debarred from a career in business management. As a start Harrow and New College, Oxford, were conventionally enough Establishment, but he then studied accountancy, and after a spell with Barclays Bank (including three years in South Africa) he joined the Board of the Employers' Liability Assurance Corporation and became its managing director in 1933. He was then 38. This was good going by any standards. Favourable winds from the Establishment may have helped, but it seemed to be common ground among his contemporaries that he was a man of lucid intelligence, agreeably supplemented by a lean and elegant presence and a courteous manner.

His new job took him a good deal to the United States, and it was probably the knowledge he thus acquired of the American business scene that paved his way for the appointment as Governor of Bermuda during the early years of the war. He remained in Bermuda until 1943

2

and the Government then invited him to take over the full-time chairmanship of BOAC.

Already assumptions were being made about eventual victory and the problems that would then arise, and as chairman of BOAC Knollys had primarily to lay plans for the post-war period. He stayed in the appointment until 1947, and then returned to the Employers' Liability Assurance Corporation.

At this time Sir Ronald Weeks was emerging as the central figure at Vickers. He had joined the Board in 1945 with a glowing reputation for drive and organisational efficiency earned as Deputy Chief of the Imperial General Staff, an appointment created in 1942 to enable him to take responsibility for army organisation and equipment. In 1949 he became Chairman of Vickers, and held office for seven years – until, in fact, he was succeeded by Knollys. It was he, exercising one of the Chairman's prime duties – to find and groom a successor – who brought Knollys into Vickers, first as a Director in 1952 and then Deputy Chairman in 1955.

When he succeeded to the chairmanship in 1956 Knollys was only the eighth Vickers Chairman since its incorporation as a public company in 1867, and only the fifth 'outsider'. From 1873 to 1926 the chairmanship had been held by a member of the Vickers family, and the family had, of course, dominated the business from its foundation in Sheffield in 1827 to its incorporation in 1867.

Until 1926, when Douglas Vickers retired on reaching the age of 65, members of the Board had always been full-time executives. With the appointment as Chairman of General Sir Herbert Lawrence, not only were non-executive directors introduced (including prominent public figures such as Sir John Anderson), but executive membership became restricted to the chief executives of the main businesses in the Group and the Director of Finance.

The degree of dominance achieved by the Chairman depended to a large extent – as it does between a Prime Minister and his Ministers – on his own will, capabilities and strength of personality as against the will, capabilities and strength of personality of the men at the top of the operating groups. Lawrence had been a strong Chairman, but the tune was really called by Charles Craven, an extremely able and forceful chief executive. With Weeks, however, there could be no doubt about the Chairman's dominance.

The Times had this to say:

Weeks made the position of chairman of Vickers, the parent of the diversified group comprising engineering, shipbuilding, steel and aviation, an executive one, partly because he realised the need for firm direction at the centre, partly because of the untimely deaths of two of the most senior executives in the group within a year of each other,[*] and partly for personal predilection. His varied experience, his knowledge of the world and of people in government and the rest of the business world, his strong and resolute personality, his human understanding, fitted him admirably for the position. As a result, when he retired from the chairmanship in May, 1956, Vickers was a thriving and well-knit organisation, worthy of its distinguished past.

Vickers was, in fact, a federation of powerful kingdoms led by strong-willed men of proved professional competence. From 1927, when the merger took place between Vickers and Armstrongs, the most powerful kingdom had been Vickers-Armstrongs Limited, itself containing three kingdoms—those of the shipbuilders, the engineers and the aircraft makers. So large and important were these three businesses that in 1955 it had been decided to form them into separate companies, each with its own executive board. This had the advantage, not least, of enhancing the status of the most senior managers.

A lesser but significant Vickers-Armstrongs company was also formed to manage the tractor project so that, in all, there were four Vickers-Armstrongs companies.

Apart from these companies the main subsidiary kingdom in 1956 was English Steel Corporation (recently returned to private ownership as a result of the 1951 General Election, with Vickers holding 75 per cent and Cammell Laird 25 per cent). In Australia an overseas kingdom existed, with plans for its expansion, and negotiations were in train to set up a Canadian kingdom by re-acquisition of Canadian Vickers (created by Vickers in 1911, but sold to Canadian interests in 1927).

The Vickers empire thus contained many kingdoms, widely diverse in products and location, and the men leading them inevitably thought first and foremost of the strength and success of their own activities. They competed for available resources and chafed against the pull of reins from Vickers House. Traditionally this was particularly true of the shipbuilders, and at Barrow it was further intensified by the fierce local pride of all Barrovians. From their isolated position on the Furness peninsula, facing southwards across Morecambe Bay, they regarded all non-Barrovians with suspicion and took deep pride in their own down-to-earth competence and self-sufficiency. Barrow was Vickers and, for them, Vickers was Barrow—a belief fostered by the frequency with

[*]Sir Robert Micklem, Chairman of Vickers-Armstrongs Limited, and Sir Hew Kilner, in charge of aviation interests.

which Barrovians who moved to other parts of the Group seemed to rise to the top. At Vickers House in London it was a stock joke that to go to Barrow you needed your passport and a strictly 'this-visit-only' visa. A Chairman of Vickers visiting Barrow was in a position similar to that of the monarch visiting the City of London when the hilt of the City Sword is offered, touched and returned to the Lord Mayor.

The need to hold the ring explained in part the policy of importing 'outsiders' to take the chairmanship, but no less important was the need for a Vickers voice to be heard and listened to with respect in the highest reaches of Whitehall and Westminster. For this role both Weeks and Knollys were admirably equipped. Of Knollys a Cabinet Minister had said 'Edgey would charm the fruit off any tree.'

Though in terms of numbers employed and of capital employed Vickers could be excelled by other British companies, it had established a position of unique national significance. Not only was it engaged in four basic manufacturing industries – shipbuilding, engineering, aircraft and steel – but from the 1890s, when it moved into naval shipbuilding, it had come to be at the very heart of British defence programmes.

It is almost certainly no exaggeration to say that without Vickers-Armstrongs the Second World War could not have been won. Statistics speak for themselves. The Spitfire and the Wellington both came from the Vickers stable. In *Jane's 100 Significant Aircraft* the Spitfire is described as 'the most famous fighting aeroplane of all time'. The Wellington, which was the mainstay of British bombing operations during the early years of the war, was constructed on the geodetic principle, first used in another Vickers aircraft, the Wellesley, creator of a long-distance world record in 1938 in a flight from Egypt to Australia. By the end of the war Vickers had been responsible for the production of nearly 20,000 Spitfires and Seafires and over 11,000 Wellingtons. Barnes Wallis, a Vickers man, had invented the bouncing bomb that breached the Monhe and Eider dams in 1943, and followed it with deep penetration bombs, Tallboy and Grand Slam, to wreak havoc with the E-boat pens, the V.1 launching sites, the 'Tirpitz' and enemy communication systems. In the production of these bombs other Vickers men, and several Vickers works, had a central role.

From the Vickers shipyards at Barrow and Newcastle came a rapid flow of submarines and destroyers, a new era of smaller ships to succeed the aircraft carriers like *Illustrious*, *Indomitable* and *Victorious* and the big-gun ships like *King George V*, all Vickers built. Again statistics tell the story. During the period of hostilities the two Vickers shipyards, at Barrow and Newcastle, delivered one battleship, five cruisers, nine aircraft carriers, one monitor, 22 destroyers, 146 submarines, 60 landing craft and barges and nine escort vessels, and in addition delivered 10 merchant ships and a transport ferry.

From the Engineers came tanks by the thousands together with a vast range of guns for both the Army and the Navy. In the early part of the war, and especially during the period when Britain stood virtually alone, Vickers' ability to move quickly into mass production had been vitally important: from the outbreak of war in September 1939 to the end of 1940 the Crayford Works alone produced nearly 20,000 machine and gas-operated guns, about half the national production, and in field artillery Vickers as a whole supplied some two-thirds of the national output during the same period.

The armament capacity available at Vickers was therefore a matter of prime national importance, and Vickers had some right to expect that the Government would have regard to the Company's problems when, almost at the proverbial stroke, armament demands fell from the insatiable to the negligible. Confronted with this situation at the end of the First World War, Vickers had been left to cope as best it could, involving frantic attempts to fill empty capacity with products as far removed from its experience as sewing machines and motorcycle engines – products which might serve as an immediate palliative, but offered no prospect of long-term viability and carried a legacy of loss and anxiety.

At the end of the second war the Government again exercised the break clause in its major armament contracts, but at the same time it declared a wish to disarm by stages and asserted its interest in the development of new weapons based on the striking advances made in weapon technology during the latter stages of the war. This assertion was accompanied by a 'request' that Vickers should maintain capacity for certain types of armaments, for example torpedo capacity at Weymouth Works.

During the winding-down period, Government also provided Vickers with a contract to manage a project to use the aircraft shadow factories at Chester and Blackpool for building pre-fabricated housing units. More importantly, it asked the Company to produce tractors for use in the groundnut scheme in East Africa. Government enthusiasm for this scheme was strongly linked with its desire to have available a heavy tractor of British design and manufacture, partly because of the shortage of dollars to buy American tractors. Vickers accordingly found themselves under heavy pressure to push ahead as quickly as possible with a tractor project, the inducement being a projected 'immediate' order rate of 500 tractors per annum for the groundnut scheme, rising to 1,000. As a stop-gap arrangement Vickers undertook to convert a number of Sherman tanks for use as heavy tractors, and |these Shermans were delivered to Elswick, reported one observer, 'complete with desert sand and Hersche chocolate bars in the lockers'. In the event, not many 'Shervicks' had been produced when the groundnut scheme collapsed. By then, however, Vickers had become so involved in the design and development of a tractor that it saw little alternative but to carry on. In any case, Vickers needed a new peace-time engineering product and the tractor seemed to offer a vast potential. This potential was never to be realised, and for well over a decade the tractor project hung round Vickers' neck like the proverbial albatross before the corpse was expensively interred. At the time, however, it provided what seemed a sensible use of resources no longer required for defence production.

A stronger place in the printing machinery industry and in the bottling machinery industry was sought in 1947 by two small but important acquisitions – of George Mann & Company Limited, a Leeds firm specialising in offset litho presses, and of G. J. Worssam & Son, manufacturers of filling machines and other equipment for breweries, a business nicely complementary to that of Robert Boby, already in the Vickers camp.

With Manns Vickers already had a close link through an agreement for the manufacture at Elswick of part of the Mann order book, then heavily loaded by post-war demand. Elswick had also taken over a licence agreement with American Type Founders, which it reinforced later with a licence to manufacture Scott newspaper presses.

American licences were also taken in an effort to fill other denuded capacity, notably for the manufacture of power presses. Here again action to meet urgent needs carried long-term disadvantage for the engineering division. Instead of developing their own commercial products with sufficient determination the Engineers had, however un-willingly, fallen back into the role of operating jobbing activities – filling capacity until more armaments orders turned up. When the Korean War broke out in 1950 this view must have seemed well justified for once again, almost overnight, the Government called for urgent action to meet its defence requirements. The engineering division's results improved accordingly, but a high price was to be paid in the longer run.

The aircraft division had in the meantime made a successful transition from war to peace. Even before the end of the war it had begun work on a twin-engined civil airliner, the Viking, developed from the Wellington, as an aircraft suitable to meet the immediate operating needs of the civil airlines. The Viking served this purpose admirably. The first machine flew in June 1945, and by 1948, when manufacture ceased, 163 had been produced. British European Airways took a substantial number, but many went to foreign airlines, and 12 to Transport Command, including four for the Royal Flight.

Vickers did not see the Viking as more than a stop-gap, however, and at Weybridge George Edwards, the new Chief Designer in succession to Rex Pierson, was at work with his team on the world's first turbo-propeller airliner. This was to be the Viscount which made its first flight in July 1948. The future of the aircraft seemed to be in doubt when BEA at first showed a preference for a piston-engined machine, but the clear merits of the Viscount, coupled with the enthusiasm of overseas buyers, vigorously encouraged, transformed the situation. The transformation was further aided by a Government contribution of £1.5m towards the completion of development, a contribution which in due course brought a return of £3m in royalties. By 1956 the Viscount was established as an outstanding success, not only as an engineering achievement, but also as a money-maker, both for Vickers and for its operators.

Moreover, military aircraft were simultaneously being developed successfully, notably the Valetta (a version of the Viking), the Varsity (a highly efficient 'flying classroom'), and the Valiant, first of the 'V' bombers, from which the first British atomic and hydrogen bombs were dropped during the development tests in the Pacific in 1956 and 1957.

All in all, the aircraft division had come through the first post-war decade remarkably well, and the shipbuilders, too, had done considerably better than might have been expected – a fair reward for the heavy expenditure undertaken in modernising the Yards at Barrow and Newcastle. If there had been a sharp decline in naval contracts, the shipping companies went far to filling the gap with orders to restore their diminished fleets. Tankers especially were in demand, and by 1956 Vickers had also built the great liners *Himalaya*, *Chusan*, *Orcades*, *Oronsay* and *Orsova*.

Some naval orders continued, chiefly for submarines, reflecting Vickers' central position as builders of the Royal Navy's submarines which had begun with the first *Holland* in 1901. Particularly significant were the experimental submarines *Explorer* and *Excalibur*, based on German experiments late in the war with propulsion by use of hydrogen peroxide. Their significance was to lie more in their limitations than in their success. Though no submarines had ever before travelled as fast under water, they could sustain these speeds in only short bursts, and it was established that the future lay rather in propulsion by nuclear energy. The HTP team at Barrow was switched accordingly, and Vickers participated from the start in the development of nuclear submarines for the Royal Navy.

For steel it had been an adventurous ten years. With the Labour Party coming into power in 1945 it was inevitable that English Steel Corporation would be nationalised, though Vickers argued that the works at Sheffield was less engaged in the production of bulk steel than in the production of ingots for their own engineering activities. When in due course nationalisation did take place Vickers emerged with a net profit of £7.7m on the book value of their holding to set against the accumulated profits they had ploughed back into English Steel. Two years later, in 1951, the new Conservative Government set about the process of de-nationalisation. Not without some misgiving, for iron and steel were clearly destined to be a political shuttlecock, Vickers and their partners, Cammell Laird, put in a bid for the return of the business on a 75:25 basis. It cost £10m cash for the equity and £820,000 for the debentures. The negotiations lasted until June 1954, and so Vickers had to pick up the threads of ownership again after three years. It did so in a bullish mood by planning to build a very large melting and rolling mill plant on a disused colliery site of 500 acres at Tinsley Park in Sheffield.

The return to the fold of English Steel Corporation had the effect of boosting Group profits considerably in 1955. Total pre-tax profits for the year were £11,981,515. Towards this total ESC contributed £4,480,000, some 37 per cent. The three Vickers-Armstrongs companies jointly put in £6,515,000, or about 54 per cent. At that time no breakdown was published of results from the three companies: they were, in fact, £2,757,000 from engineers, £2,699,000 from aircraft and £1,059,000 from shipbuilders. Other contributions to profit in 1955 came from the Powers-Samas Accounting Machines Group (£641,000), Cooke, Troughton & Simms (£216,000), Vickers Ruwolt in Australia (£209,000), and Robert Boby Limited (£140,000).

Vickers had acquired a controlling interest in Powers-Samas in 1948 in the expectation that punch-card tabulating machines would fill a gap in capacity at the Crayford and Dartford Works caused by reduced defence requirements for gunnery control equipment. This purpose was largely achieved in the short term and in 1955 Vickers bought the remaining shares, to make the total cost of the acquisition £9.5m. In fact, the business had a national significance perhaps not then fully realised. By a process of mergers it was to lead to the formation, first, of International Computers and Tabulators Limited (ICT), in which

Vickers had the biggest single holding, and then to International Computers Limited (ICL), the major British computer-machinery company (from which Vickers withdrew in 1970 by sale of its holding).

Cooke, Troughton & Simms Limited had been in the Vickers fold for a considerably longer period. It reflected Vickers' interest in scientific optical instruments, in particular range-finding and fire-control equipment, and represented the putting together of two acquisitions, of Thomas Cooke & Sons, of York, in 1915 and Troughton & Simms Limited, of London, in 1921. With Cookes Vickers had been doing business as far back as 1896. Both firms had been long established – Cookes from 1837 and Troughton & Simms from 1826 – and each had a high reputation in the industry so that their names were retained in the title of the merged company (only later, in 1963, was the title changed to Vickers Instruments). Though not destined to be a major Vickers business, it is still a significant activity, and has consistently returned profits in the £200,000–£300,000 range.

The background to Robert Boby Limited was not dissimilar, its full acquisition in 1930 arising from work done for Vickers on armament sub-contracts, including shells and paravanes, though its main business was (and continued to be) the manufacture of malting and similar equipment at its works at Bury St Edmunds (where it had been founded by Robert Boby, an ironmonger, early in the 19th century).

The profit of £209,000 from Vickers Ruwolt in 1955 was the lowest from the Australian company since its acquisition in 1948 (in 1953 profit had totalled over £400,000). This falling away was attributed to intensified competition, though no second thoughts were considered necessary about plans to acquire the heavy engineering business of C. & A. Hoskins in Western Australia. The Hoskins acquisition was completed in 1956 so that Vickers then had interests in Melbourne (Ruwolts), Perth (Hoskins) and Sydney (the shipbuilding, shiprepairing and engineering business carried out on Cockatoo Island in Sydney Harbour under a leasing agreement with the Australian Commonwealth Government).

One Vickers company not showing a profit in 1955 was George Mann & Co. The loss was only £31,000, but it followed a loss of £89,000 in 1954, and was deeply disappointing given the high expectations that had been attached to its acquisition as a means of giving Vickers a stronger footing in the printing machinery industry.

There could be no doubt, however, that overall 1955 had been a very good year for Vickers. Profit figures were the best on record. Because of heavy cash needs there had been in 1954 a Rights issue of Ordinary Shares and a new issue of £6m 4 per cent Unsecured Loan Stock. The

Rights issues and a previous scrip issue had more than doubled the value of the issued capital in Ordinary Shares. Though the dividend for 1955 was fixed at 10 per cent this, in fact, represented an increase over the 15 per cent declared in 1952 and 1953. The Stock Market took a buoyant view of the Company's prospects – the dividend was covered some 2½ times – and the middle share price on 1st June 1956, the day after the annual general meeting, stood at 40/6d.

Certainly Lord Knollys could have had little reason on that day – his first as Chairman – to dispute *The Times'* view that Lord Weeks* had handed over to him 'a thriving and well-knit organisation, worthy of its distinguished past'.

Equally, however, he could have had little doubt that serious problems lay ahead and that far-reaching decisions would have to be taken before long about the future shape of the company. As a broad objective it had already been agreed that the company should seek to make itself less dependent on armaments and on Government contracts, but moves in this direction had been hesitant, and armament orders arising from the Korean War had blurred the issue. The new interests in Canada, Australia and India had been undertaken more from Lord Weeks' conviction that great industrial companies should play their part in strengthening Commonwealth links than in pursuance of a policy of commercial diversification. In the United Kingdom the attempts to introduce commercial products had met with mixed fortunes, and the tractor project in particular was draining resources without much prospect of early reward. Here was the hub of the matter. Any moves on a large scale into new commercial products must make heavy financial demands, whether for acquisitions, for capital expenditure or for research and development, and the Company was finding it increasingly difficult to marshal adequate resources, especially against the background of escalating costs in aircraft development. Profits from the Viscount and the Valiant were beginning to disappear in the cost of developing Vanguard, and it was clear that the Company would be unable to carry alone the burden of financing new civil aircraft.

Though it might be unreasonable to expect the Chairman to produce all the solutions to these problems out of his own personal hat, prime responsibility nevertheless lay with him to identify overall priorities, to see that the issues were thoroughly analysed and debated and then to guide the Board to sensible decisions. To this extent the way he performed his task was highly important to the 90,000 people employed by the Company and the 62,000 people and scores of organisations who bought its stock.

* Created Baron, 1956.

Part Two
1956–1962

I

Change and Venture

THE KNOLLYS CHAIRMANSHIP WAS TO LAST FOR SIX YEARS, UNTIL
1962.

His style of chairmanship was markedly different from that of Lord
Weeks, inevitably so given the different temperaments and backgrounds
of the two men. Weeks was both a notable autocrat and a notable
extrovert. He enjoyed people and mixed easily. He liked visiting the
Company's works and tried to ensure that the Board met at a Works
once or twice a year. On those occasions most of the senior managers
would probably have been inveigled into sessions of *chemin de fer* far
into the night. He warmly admired the test pilots of the aircraft division
and established himself on Christian-name terms with them. His ability
to mix with hard-headed engineering bosses was based on more than a
cheerful gregariousness, however. The world tended to think of him as
a soldier, which was not surprising given his decorations from the
First World War, his achievements as a Deputy Chief of Staff in the
Second World War and his continued use of the title of Lieutenant-
General. In fact, his father was a mining engineer and he was brought
up in an industrial environment. He had gone to Pilkington's, the glass
manufacturers, as a technical trainee in 1912 and between the wars
returned to become a works manager and later chairman of the executive directors of the company. He therefore had a first-hand experience
of problems at the shop-floor level, and could talk to Vickers works
managers with knowledge and understanding of their outlook and
problems.

Knollys did not have this advantage of experience in industrial
management. It could not be expected that he would find it as easy or
as rewarding to spend time visiting the works. Nor temperamentally
did he mix as easily at the works levels as Weeks had done. Without
remaining aloof, for he was always approachable and invariably courteous, he nevertheless stood somewhat apart, preferring to move in the
highest echelons of policy-making and negotiation where his air of
natural authority seemed rightly to place him. With him went that aura

of 'amiable omnipotence' which has been attributed to those born to the purple.

He also believed that one's affairs should be conducted with style and elegance. 'You will find you have an expensive chairman', a Vickers director was told. Certainly one of the first matters to which he directed his attention was the finding of headquarters for the Company less undistinguished and shabby than the block of offices it occupied in Broadway (part of the site now occupied by New Scotland Yard). Vickers thus became involved in the building of what was to become Millbank Tower, on a splendid site on the north bank of the Thames, between Lambeth Bridge and Vauxhall Bridge. In fact, by the time it had been completed Knollys had ceased to be Chairman, though he was still a member of the Board, and the opening ceremony, in May 1963, was performed by Lady Knollys. The move was far from being welcomed by all in Vickers, and voices could later be heard suggesting that the Company's period of travail dated from then: but at least there could be little doubt that Vickers had acquired control of a building of exceptional architectural merit on a site which ensured its recognition as a London landmark.

Another matter in which the new Chairman quickly found himself immersed was the selection of new members of the Board. The Company's established practice was for executive members to retire at 65 and non-executives at 70 (assuming that they chose not to retire before then). Several were approaching this age and new blood would be required.

There was also the question of the succession. Here action was taken within a few months by appointing Charles Dunphie as Managing Director. Like Weeks, Charles Dunphie was a General and a man with a distinguished war record, but, unlike Weeks, he had been a professional soldier. His war service took him to France and to North Africa, where he commanded the 26th Armoured Brigade. At the end of the war he was Director-General, Armoured Fighting Vehicles, in the Ministry of Supply, and in that appointment both knew Vickers and was known by Vickers. What Vickers knew they liked and in 1948 Dunphie came to the engineering side of Vickers-Armstrongs with the intention that he should in due course become the V-A Managing Director. To acquire experience in running a commercial enterprise he first undertook a number of tasks in the engineering businesses, in particular as managing director of the Robert Boby subsidiary at Bury St Edmunds. In 1952 he had duly been appointed Managing Director of Vickers-Armstrongs Limited with a seat on the Board of Vickers Limited – perhaps a little earlier than expected because of the untimely

deaths of Sir Robert Micklem and Sir Hew Kilner. Though a professional soldier for 27 years, his family background was remarkably
varied and made it unlikely that he would lack flexibility in ideas and
attitudes. His father, Sir Alfred Dunphie, had been called to the bar but
made banking his career, becoming a director of Coutts & Company,
and also serving for 25 years as Assistant Treasurer to Queen Alexandra.
His mother was a surgeon's daughter. Like his father, Charles Dunphie
held a Court appointment, being a member of the Honourable Corps
of Gentleman-at-Arms. His grandfather, Charles James Dunphie, was
art and dramatic critic of the *Morning Post* for over fifty years. Of
Charles James Dunphie, the *Dictionary of National Biography* says: 'He
was a graceful writer of Latin, Greek and English verse and a semi-
cynical essayist'. It also describes him as 'of handsome appearance and
polished manners'. This description could equally have been written of
his grandson. In Vickers Charles Dunphie was noted for the consideration he showed to those around him. His appointment as Managing
Director of Vickers Limited confirmed his role as an *alter ego* to the
Chairman, a role he had already assumed in effect during the Weeks
chairmanship when he became, first, Managing Director and then
Chairman of Vickers-Armstrongs Limited.

Two other appointments followed which were to have great significance. In February 1957 Eric Faulkner, then Managing Director of Glyn
Mills & Company, joined the Board as a non-executive, and at the end
of 1958 Sir Leslie Rowan came to Vickers as Director of Finance.

Glyn Mills had been Vickers' main bankers since 1867, when the
Company was incorporated, and there was a traditionally close relationship between those working at the top of the two organisations. Eric
Faulkner, then 43, was widely recognised as one of the brightest lights
in the world of clearing banking, and later he was to shine even more
brightly as Chairman of Lloyd's Bank and a leading figure in the life
of the City. Apart from the authority which he thus brought to the
Vickers Board, he took a very positive interest in the affairs of the
Company and came to exercise a crucial influence in the Board's
discussions.

The non-executive directors as a group provided a valuable range of
professionalism and experience. Two of them—Terence Maxwell and
David Pollock—remained members for almost the whole period
covered in this book. Indeed, by the time he retired, Colonel Maxwell
had been a member of the Board for no less than 41 years, having been
appointed as a very young man in 1934. His exceptional drive and
determination were already clear by then. After becoming a barrister
he had elected to make a career in banking, and he was closely involved

in the negotiations with Armstrong Whitworth that resulted in the great merger in 1927. He could also contribute a considerable knowledge of the political scene. He sat as a Member of Parliament during the 1930s and his father-in-law was Sir Austen Chamberlain. For periods of his long service on the Vickers Board he was as much executive as non-executive since he took particular responsibility for Powers-Samas after its acquisition and was in due course Chairman of ICT. His agile mind and quick tongue, allied with a commanding physical presence, made him a very forceful member of the Board.

So, too, though in a quieter style, a significant influence was wielded by David Pollock, a solicitor who had moved into top business management with the Pearson empire, headed by Lord Cowdray. He was brought to the Vickers Board by Lord Weeks, who had been greatly impressed by a paper he had done on the provision of aid for airline operators (a paper which led to the formation of a consortium of bankers who purchased aircraft and leased them to operators). Sitting next to Weeks at the board table of the Westminster Bank was not, in fact, Pollock's first opportunity to be made aware of Vickers: as a Commander RNVR his war service had been almost entirely on Vickers-built ships, notably *King George V* and *Illustrious*. Not least, he recalled a telegram sent from Malta to Vickers by the captain of *Illustrious* after she had survived very heavy damage by aircraft attack. It said: 'I am very proud of your ship.'

However valuable the role of well chosen non-executive directors, it is with the executive directors that the main responsibility lies for conduct of the Company's affairs. The appointment of Sir Leslie Rowan as Director of Finance was therefore a matter of cardinal importance. It must be assumed that Lord Knollys foresaw the possibility that Leslie Rowan could become the next-but-one chairman, and was therefore ensuring the succession twice-over, but no one could be sure how a distinguished senior civil servant, a Treasury knight indeed, would perform in industrial management. Still only 50, Rowan had behind him a career of unchecked brilliance – academically at Cambridge, hockey captain of the university and of England, rapid promotion in the Administrative Civil Service, Principal Private Secretary to Winston Churchill and Clement Attlee at No 10, Economic Minister at the Washington Embassy and then, at the time of his retirement from the Civil Service, Second Secretary at the Treasury. No doubt there were several factors which led to his decision to leave the Civil Service, but he made no secret of the revulsion he felt about the handling of the Suez situation in 1956. Certainly he was a man inclined to react emotionally to events, reflecting perhaps his family background for his father and mother had been medical missionaries in India. He was also a man believing passionately in the virtues of team work and loyalty to the cause and the leadership.

These feelings it was always probable would be put under strain in an organisation such as Vickers. Pride in being 'a Vickers man' was unlimited, and long service was not least of the hallmarks of 'a Vickers man', but nevertheless the pride was likely to be focused on the particular segment of Vickers activities in which a man was engaged rather than on the overall concept of a conglomerate group. Sitting in Vickers House you might feel that it was one-for-all and all-for-one, but out in the works the feeling was more likely to be 'them' and 'us', with 'them' seen as an expensive overhead of questionable usefulness.

At the time of Leslie Rowan's appointment changes were also taking place, or becoming imminent, at the head of the operating groups. George Edwards, still a relatively young man, was firmly established in charge of the aircraft company and had become a member of the Vickers Limited Board in 1955: his powerful and engaging personality would have made him a key figure in any case, but he was also emerging as a figure of national eminence because of his central role in the aircraft industry and he had received a knighthood in the 1956 New Year Honours. In the engineering and shipbuilding companies, however, the leadership situation was less clear-cut and new men had to be brought forward. For the shipbuilders the man to emerge as Managing Director in 1957 was George Houlden, a Scot educated and apprenticed in Aberdeen, who had been with Vickers since 1924, serving at both Barrow and Naval Yard, Newcastle. As well as being a highly competent professional he was a friendly man, but he nevertheless saw himself, first and foremost, as custodian of the interests of the shipbuilders and at Board meetings liked to play the shipbuilding hand close to his chest. He was said to support his apparent taciturnity by skilful use of a hearing aid, which could be persuaded to issue piercing whistles at difficult moments in discussion.

The man at the top of the engineers, Peter Muirhead, was also a long-service Vickers man, but his health was fading, and in January 1958, though remaining a member of the Board, he was succeeded as Managing Director of the engineering company by W. D. ('Bill') Opher. A few months later Muirhead died, and Opher became first a Special Director, and then, in 1959, a member of the Board. He, too, had been with Vickers for many years, his service including a stint in Newcastle which helped to offset the disadvantages – in the Vickers engineering environment – of having been born a Londoner and apprenticed to a firm of London engineers. Shrewd and tough, but also

3

highly gregarious, he had the ability to look beyond the boundaries of the engineering company to the wider interests of the Vickers Group, and so to play rather more than a 'representative' role on the Board. This was not always easy for an executive from an operating business, though anyone appointed to the Board was in theory there for his individual qualities with a requirement to act in the interests of the stockholders in managing the Group as a whole.

There was thus no prescriptive right for the chief executives of the major businesses to be members of the Board, though in practice this usually happened. In the instance of English Steel Corporation, its managing director in 1956, Frederick Pickworth (later Sir Frederick) was a Vickers director. On his death in 1959, however, Lord Knollys took over the chairmanship of the Corporation and it was not until 1962 that the managing director, Dr W. D. Pugh, was appointed to the Vickers Board.

One other Board appointment during the Knollys era should be noted because of its later significance. R. P. H. Yapp became first a Special Director and then a member of the Board in 1958. He represented the third generation of his family to serve Vickers, and his father, Sir Frederick Yapp, had been Managing Director of Vickers-Armstrongs Ltd, with a seat on the main Vickers Board, during the war period. His own career had begun in the City, but he joined Vickers in 1931 and at the time of his appointment to the Board was Director of Administration. His appointment was welcomed, not only because of his shrewd and patient thoroughness, but also because of his talent for friendliness, reinforced by a twinkling sense of humour and the ability to produce an incisive *bon mot* at critical moments. Such was his integrity that, when he was Director of Contracts, it was said in the Admiralty: 'We would always accept his word implicitly'. Some ten years later he was to become Managing Director.

The making of appointments is, of course, a principal duty of the Chairman, and much depends on his skill in putting the right man in the right place. In this respect he has a considerable power to command events.

In other respects, however, a Vickers chairman, no less than a Prime Minister, has quickly to learn that you must spend much of your time responding to events rather than commanding them. Also that the pressures are not only external, but internal. A Prime Minister is under pressure from his Ministers. A Vickers chairman is under pressure from the chief executives of the operating groups. He is in the middle of the sandwich. The Government seeks to control and change the economic and industrial environment in which the company operates, and the

men at the top of the operating groups seek to build up the momentum
of their own businesses to the point at which it is difficult to resist.

A Conservative Government was in power throughout the six years
Knollys was Chairman, and to the extent that this disposed of any
immediate fears about nationalisation he was fortunate. But the Macmillan administration sought to develop very positive policies for
re-organising the aircraft industry and for re-shaping defence policies.

Both had traumatic effects for Vickers. Within four years Vickers
had ceased to be a direct manufacturer of aircraft, and its aircraft
interests had become a 40 per cent holding in the British Aircraft
Corporation. In defence, the apparent implications of the Sandys White
Paper in 1957 caused Vickers to conclude that, except possibly in naval
shipbuilding, its role as a supplier of defence equipment must rapidly
diminish. This conclusion was to prove mistaken (some conclusions of
the White Paper having themselves been totally misleading), but in the
long run it probably did Vickers more good than harm by intensifying
the Company's consideration of its own future, and by accentuating
the importance of diversification into commercial products and
commercial attitudes as quickly as possible.

In steel, during Knollys' chairmanship, the Tinsley Park project took
the centre of the stage. It provided for building a new works, containing
melting and rolling mill plant, on a disused colliery site in Sheffield.
Its escalating cost caused worry and by the time it had been fully
completed in 1963 the bill had reached £26m. Finance for the project
came to a large extent from English Steel Corporation's own credit-
worthiness with the financial institutions, but Vickers provided some
assistance in the form of further equity. Added to other demands,
particularly for aircraft (heavy losses on the Vanguard and even heavier
development costs for the VC10), the Company was moving into a
period of cash stringency which was to remain with it for most of the
decade of the 1960s.

This shortage of finance inevitably put a brake on hopes for rapid
diversification. Nevertheless, planning went ahead and opportunities
were sought.

Knollys had a strong belief in the planning function. He did not
believe that it could all be left to evolutionary processes in the operating
businesses. He argued for a strong effort at the centre in both research
and planning. One of the early acts of his chairmanship was to set up
Vickers Research, based initially at Weybridge, but subsequently
moving at a cost of £200,000 to a country mansion at Sunninghill. It
was followed by the creation of a Group Forward Development Department at Vickers House, and to run the Department Knollys brought

in J. G. Lloyd, who had successfully introduced new products at Crayford Works. The Research Establishment and the Planning Department were expected to work together in identifying new products and in finding a balance between 'own product' development and product acquisition.

Both innovations met with a degree of resistance and scepticism from the operating groups. The aircraft and shipbuilders considered themselves to be in the forefront of their own highly specialised technologies—rightly so on the evidence of performance—and could see little reason why they should be required to contribute, directly or indirectly, to the cost of a central research department. So, too, they felt more competent than anyone else to plan ahead in their own activities. The only outcome of a central planning department, as they saw it, would be to divert resources to new and speculative activities at their own expense. For English Steel Corporation much the same considerations applied, but in any case the Corporation had a well established and accepted separate identity, and was already launched on its great development project at Tinsley Park.

For the engineers it was perhaps another matter. While the other businesses were very large and homogeneous, the engineers though very large in aggregate were in reality a conglomeration of relatively small businesses, producing a considerable diversity of engineering products. To some extent the engineering company had become a rag bag of activities. This was largely for historical reasons. In the days when Vickers had been essentially an armaments company, a number of sub-contracting businesses had been acquired to ensure control over performance and delivery. In some instances the original need had disappeared, but the business remained a Vickers subsidiary and had either to be closed, sold or adapted to commercial production. Added to this, of course, was the need to find commercial products for capacity that had become surplus to armament requirements. The engineering group therefore needed to identify commercial activities with sufficient growth potential to justify substantial expenditure in building them into large or at least medium-size businesses. As compared with the other groups they also had a relatively small headquarters organisation of their own, and so tended to look to Vickers House for aid and support. For these reasons, the creation of a central research establishment and a central planning department was not without its attraction to the Engineers, especially when projects were not being directly costed to them, and as events turned out the work of both organisations became largely directed towards the needs of the engineering group.

Lord Knollys sought also to create a measure of centralisation in manpower recruitment, training and career development. In this he was expanding policies initiated by Lord Weeks who had a strong personal interest in technical training, not only for Vickers own purposes, but as a matter of national importance. Within Vickers, a Chairman's Committee on Education had been set up and a Group Education Officer (C. L. Old, Principal of the College of Technology, Wolverhampton) was appointed in 1957. Plans had also been set in train for Vickers to present a new hall of residence to Imperial College, London, with the building made available to Vickers during Easter and summer vacations for training conferences and courses. The formal ceremony of presentation of the building, and its naming as Weeks Hall, took place in September 1957.

To carry matters further Knollys set up a series of annual conferences of senior managers from all parts of the Company. The first conference took place at Cooden Beach in May 1958, and three more followed – at Vickers House in 1959, at Weeks Hall in 1960 and at Stratford-upon-Avon in 1961.

From these discussions emerged a Management Development Scheme which was introduced throughout the Group at the beginning of 1960. The scheme called for identification of younger managers considered to have potential for senior appointments so that their careers could be planned systematically, including attendance at external courses as and when this seemed desirable.

The conference also approved a scheme of scholarships designed both to attract outstanding new recruits and to provide university and technical college training for suitable young people already recruited. One innovation was the creation in 1961 of four open scholarships and four open bursaries, awarded annually, for school leavers seeking to go to university. At Cooden Beach a strong case had been made for trying to win recruits at sixth-form level rather than after graduation, so that management training could be fitted into the university years. Since very few strings were attached to the Scholarship Awards they represented bread upon the waters. The scholar could read his own choice of subjects, he was under no more than a moral obligation to join the Company on graduating, and if he did so there could be no assurance that, after further management training and experience, he would not choose to make his career elsewhere. Moreover, it was bound to be many years before the scheme could be properly evaluated. What did become quickly apparent, however, was its ability to attract outstanding boys, and the scholarship selection committee, headed by the Chairman, found itself with an embarrassment of riches, drawn from a large cross-section of public and grammar schools.

Though the conferences were concerned essentially with management recruiting, training and development, they reviewed also the whole gamut of education and training within the Company. Conferences, courses and inter-works visits were instituted for foremen, and it was agreed that facilities for apprentices, already on a large scale, should be further strengthened, with encouragement given to participation in external character-building projects such as the Outward Bound courses and the Duke of Edinburgh Awards Scheme. Most of the large works had their own apprentice training schools and at any one moment, throughout the Group, there were likely to be over 4,000 apprentices in training. A Vickers apprenticeship, the Company claimed, provided a groundwork which would go far to ensuring a young man's acceptability in any engineering or shipbuilding establishment throughout the world.

In general, the schemes and procedures resulting from the education and training conferences won the ready co-operation of the operating managements. As the number of seminars and courses multiplied, however, especially at Weeks Hall, some grumbling could be heard about the extent to which senior management time was being taken up. Concern was also expressed from the Shipbuilding Group about the psychological impact on some of the undergraduates of being given opportunities to discuss the Company's policies with the most senior directors and to criticise and argue. This, it was suggested, gave them inflated ideas of their own importance and caused a lost sense of proportion which diminished their performance in working conditions. Lord Knollys reacted sharply. A university education, he said, was designed to develop an inquiring and critical mind, and two days at Weeks Hall could not possibly be responsible for the bad effect that was alleged.

Experience with the group education and training conferences was considered sufficiently encouraging to warrant use of the technique for consideration of wider problems. Knollys had already formed a small informal group of directors to consider the future of the Company, and in 1960 he assembled at Vickers House a conference of the managing directors of the main UK operating units, together with the principal executives of Vickers Limited, some 30 in all. The purpose was to examine the overall position of the Company and to pool views on how its problems should be tackled and how its future should be shaped. An important spin-off benefit was obviously the opportunity the conference provided for the most senior managers throughout the Group to get to know each other personally and to develop the practice of talking to each other informally about mutual problems. The conference, though to some extent tentative and concerned with only the broadest issues of policy, was thought to have achieved these purposes, and paved the way for similar conferences in succeeding years. If the first two conferences necessarily dealt mainly with matters of broad policy, the later ones put a stronger focus on operational techniques in a very practical way, for example, buying and pricing policy, control of stocks and work-in-progress, work study teams, and sales and marketing, with particular reference to exports. The conferences served also as a means of reviewing the activities of the new Forward Planning Department and the central Research Organisation.

In his keynote speeches at the 1960 and 1961 conferences Knollys put heavy emphasis on the need to improve profits in order to provide the capital required for expansion. The only other source for this capital, he said, was by public issues, and these had little prospects of success, especially at a time of tightness in the money market, unless the earnings and dividend record of the Company provided clear evidence of growth potential. Dividends should be covered at least twice by earnings. He estimated that to increase the dividend from 10 per cent (at which it had become almost fixed) to $12\frac{1}{2}$ per cent, while at the same time providing a satisfactory cover, would need an increase of some £1,750,000 in after-tax profits above the level achieved in 1960, and this represented a very tall order. Projects of expansion, he pointed out, were continually coming forward, but the money would be hard to find in present circumstances, especially with approval already given for the Tinsley Park scheme put forward by English Steel Corporation.

He might have added that another continuing drain on cash resources was likely in aircraft losses and development costs. Though the Vickers aircraft business had become part of the new British Aircraft Corporation at the beginning of 1960, 'old account' projects, in particular the VC10, remained on Vickers books and represented a very heavy cash commitment—between £25m and £30m on the VC10 at its peak. Full recovery of this expenditure, even with the Government support agreed as part of the BAC agreement, would clearly be difficult and there was the prospect of major provision becoming necessary. Heavy losses had also been incurred on Vanguard (the final bill came to £16.7m).

With these factors to aggravate the shortness of cash for new projects, the conferences had as a recurring theme ways and means of tightening control of expenditure. All new schemes of capital expenditure, it was emphasised, must be examined in relation to the expected returns on capital employed. Strict criteria would also need to be applied in consideration of proposals for acquisitions; in particular it would be necessary to foresee the early achievement of a return on capital employed of between 12½ and 15 per cent before tax.

The 1961 conference discussed at some length the validity of relating profit to capital employed as the principal yardstick of performance. Progress payments from customers, it was pointed out, could have a profound effect on capital employed, and so could long-term credit given to customers. Then again, if fixed assets were valued at book values the return on capital employed would be overstated, whereas it would be understated if replacement costs were the basis of measurement. Nevertheless, in the consensus view, return on capital was so widely used as a yardstick that its use by Vickers seemed inescapable, whatever mental reservations might exist.

Knollys took a firm position. An adequate return on capital was absolutely necessary, he declared, not only on new products but on established ones. Moreover, it involved a change in the collective attitude of mind of the Vickers Group. The Group had been accustomed to rely on Government orders for armaments, and the attitude to commercial products had been that Vickers never gives up. 'This traditional attitude has resulted in avoidable losses,' he said sternly, 'and unprofitable activities should now be looked at objectively and discontinued where no prospect of a reasonable return can be seen.'

The tractor project must have been in the forefront of his mind. Now, at long last, after 13 years, a decision had been taken to wind up the tractor company. For the engineers, with their professional pride deeply involved, this was a bitter blow. In the late 1940s the call to design and build a British industrial tractor, capable of holding its own in world markets against American competition, must have made every kind of sense. Vickers had a long experience in building tracked vehicles for the Army. To adapt this experience to a heavy tractor was a challenge they could accept with total confidence. Moreover, the world market for heavy tractors in the period of post-war reconstruction and development looked almost insatiable. For Vickers this was surely the splendid new commercial product – a product of their own design and manufacture – they needed to replace armament work.

So the project was launched in 1947 on an ambitious scale, with plans later drawn up to create a new Works in Newcastle to undertake manufacture. Marketing and sales were placed in the hands of Jack Olding & Co, and Vickers acquired the Olding business a few years later. This seemed an entirely sensible decision, given that no one in Vickers at that time had any experience of this sort of selling. With hindsight however, some said that this was an initial mistake since heavy engineering products should be sold by those who make them. It was also argued that although Oldings were top of the tree so far as the UK distribution of imported tractors was concerned, they had no experience in overseas markets, where Vickers tractors had primarily to be sold. In part these arguments rested on the fact that reports from overseas agents and dealers on the experimental pre-production models gave much more encouragement than they should have done, for when the first production model came on to the market in 1953 complaints from customers poured in. Neither the transmission nor the Rolls-Royce engine were satisfactory and it was quickly apparent that the tractor would have to be re-designed and improved almost *in toto*. Some 300 tractors already sold were called in, with due financial reimbursement to the owners.

This was in itself a rude shock for the designers who now saw that the transition from armament design to commercial design contained unsuspected pitfalls. In armament design you expected to arrive at a high-quality product by a process of trial and error, but you proceeded by stages and competitive costs did not have to be the first consideration as you introduced modifications. Moreover, it was not fully appreciated that tractors, unlike tanks, were likely to find themselves in the hands of unskilled drivers, particularly in developing countries, and that care and maintenance would be much more rudimentary. The Vickers tractor was, to quote one view, 'too delicate and too expensive'. Re-design and improvement against this background inevitably took time, but considerable effort was put into the task and in 1957 a really competitive machine, the Vigor, could at last be put on the market. Even then its flexible track system reduced its usefulness in the 'pushing' role, however efficiently it might perform in pulling.

In any case, market conditions had changed. There was now, if anything, world over-production, with all that meant in terms of competitive prices, and also the wheeled tractor was beginning to take over from the tracked vehicle for pulling purposes. The Vigor performed well in an earth-moving competition during the building of the Doncaster by-pass but it had arrived on the scene too late. Nor did a smaller version, the Vikon, fulfil hopes. In 1958 the Ministry of Supply ordered 60 Vigors for the British Army, and an order for 28 came from Yugoslavia, which may have momentarily rekindled hope, but the writing was on the wall. To produce a wheeled tractor from scratch would take too long, and, though attempts were made to move into the wheeled tractor business through manufacturing licences or by acquisition, they proved abortive. Vickers-Armstrongs (Tractors) ceased manufacturing at the end of 1960. All completed tractors were sold at 50 per cent discount, generous financial arrangements were made with dealers on termination of their franchises, and arrangements set up for a continuing spares service for tractors already with customers (many Vigors were still in use a decade later). So far as customers and dealers were concerned the project thus ended in regret rather than recrimination, but Vickers were left to count the cost.

The tractor business had never looked like recovering its development and manufacturing costs. In the earlier years its costs and losses were absorbed in the accounts of Vickers-Armstrongs (Engineers), but with the formation of Vickers-Armstrongs (Tractors) in 1956 it was possible to see the extent of the annual financial drain. In 1957 the tractor company recorded a loss of £1.05m, in 1958 £1.12m, and in 1959 £0.89m. The loss for 1960 totalled £0.6m, to which had to be added a termination loss of £1.75 debited direct (less tax relief of £0.9m) to Vickers-Armstrongs general reserve. In all, the loss since commercial production began was estimated at £9.4m. More important, perhaps, was the loss of time and effort which might have been used more fruitfully in other projects. In 1960 the Engineers were still without a major commercial 'own product', and still mainly dependent on manufacturing and sales licences for commercial products and on armaments. For growth it was necessary to look to acquisitions.

Profits from Vickers-Armstrongs (Engineers) were nevertheless at a fairly satisfactory level. In the four years 1957 to 1960 the pre-tax figures were £3.45m, £3m, £3.98m and £4.58m. Only the steel makers, with annual profits steadily between £4.5m and £5m, did better among the operating groups. Against the yardstick of capital employed (at book value) the Engineers had also shown a useful improvement from 8.8 per cent in 1957 to 12 per cent in 1960, though to the extent that comparisons could be struck this was still below the national average for mechanical engineering.

The main centres of Vickers engineering activity were at Barrow, Newcastle and Crayford.

Production at the big engineering works at Barrow was largely geared to the requirements of the shipyard. The Barrow Engineers did most of the machinery installation and fitting out on Vickers-built ships. They also had a large and important 'traditional' activity in naval gun mountings and now, additionally, they had begun to design and manufacture missile launchers. Further reaching into the future, they had become closely involved in the application of nuclear energy to marine propulsion. In practice this meant contributing to the design and construction of nuclear power plant for use in submarines, a project undertaken at the behest of the Government by a new company formed in 1956 by Vickers, Rolls-Royce and Foster Wheeler. In parallel with this project they had received the main contract for building at Dounreay, in Scotland, the facility needed to test the new nuclear propulsion machinery in simulated submarine conditions. The Engineers at Barrow thus continued to have a strong base of armament work. Their main commercial work, in addition to that involved in commercial shipbuilding, lay in cement plants, pumps and diesel engines, as applied to both marine and land use. The diesel capacity acquired an extra dimension in 1957 with the award of the first of a sequence of very large contracts by the British Transport Commission for engines for the hundreds of locomotives required in the modernisation of British Rail. So considerable were these requirements, coupled with export orders, that a large shop was allocated and equipped at Barrow especially for the purpose, and by the time the orders had been completed, no less than 1,500 engines of Sulzer design had been built and installed.

At Newcastle the principal Vickers-Armstrongs engineering works was at Elswick, with Scotswood works, a little further along the north bank of the Tyne, about to provide additional capacity following closure of the tractor project. Elswick had long been the home of the armoured fighting vehicles division, design parents of the Centurion tank with which the British Army had been equipped since the war and for which large export orders had been obtained. Its commercial activities included a small but valuable business in non-ferrous metals, the manufacture under licence of Scott-Vickers presses for the newspaper industry, Clearing presses (also under American licence) and dies for the motor industry, and rolling mill equipment for the steel industry.

Crayford works also had its element of armament work, though on a lesser scale than at Barrow and Newcastle, its expertise being in fire control and aiming equipment for both ship and aircraft use. Several commercial products were being manufactured under licence – for example, petrol pumps and packaging machinery – and Crayford was also the home of the bottling machinery business developed from the acquisition of Worssams in 1947.

If capacity at Barrow, Newcastle and Crayford was reasonably well occupied, the Engineers were faced in 1960 with an urgent need to find products for the works at South Marston, near Swindon, which they had taken over from the aircraft company.

South Marston had been built in 1940-41 by the Ministry of Aircraft Production as an aircraft factory, complete with its own airfield, and was placed under Vickers control in 1942, to cope with the imperative demand for Spitfires. After the war it was acquired by Vickers and continued as a production centre for the Supermarine arm of Vickers-Armstrongs, first with Attackers and Swifts, and then Scimitars for the

Royal Navy, at the centre of its activities. When the Scimitar contract had been completed in 1960, and with no prospect of further naval or RAF orders, South Marston was left with only a certain amount of aircraft sub-contracting work, and its future lay in the balance. The task of finding replacement work (and doing it very quickly) fell mainly on S. P. Woodley, who had acquired a considerable reputation with Supermarine for drive and energy. In the light of subsequent developments he may have tried some long shots, but South Marston remained alive if not financially well. In addition to original work on plant for the accelerated freeze drying of selected foods, it undertook early development of hovercraft, designed and manufactured equipment for postal mechanisation, and moved into nuclear engineering on a modest scale with reactors designed for training and research. It also became the manufacturing location for the first product to emerge from Vickers Research Establishment – a medical linear accelerator – and later it moved into medical engineering of a different kind with manufacture of incubators and other oxygen therapy equipment developed by the Research Establishment. Circumstances had thus caused South Marston to become the test bed for a number of new 'own products' without a really adequate base load of established products.

Two other new opportunities presented themselves to the Engineers as the result of a visit to Moscow in 1958 by a team led by Robert Wonfor, then Sales Director. This visit followed one already made to China as part of a systematic examination of the opportunities available in overseas markets, particularly behind the Iron Curtain, for heavy engineering plant. Soviet interest in building chemical processing plants was already known, and since this type of plant included heavy components well suited for Vickers manufacture, the mission had these possibilities strongly, though not exclusively, in mind.

The Soviet authorities were, in fact, talking to Hans J. Zimmer, a young German physicist who had established a reputation for inventive genius in the chemical engineering industry, and had set up a sophisticated business for designing and installing plant for manufacturing synthetic fibres and rubber. The business was based in Frankfurt and employed some 1,500 people. The Russians admired Zimmer's abilities, but felt that his company had insufficient resources and backing to tackle the size of projects they had in mind, not least in the provision of credit terms. Why not come to some form of arrangement with Herr Zimmer, they suggested to Robert Wonfer, which would give Zimmer the backing he needed and Vickers the opportunity to manufacture chemical plant. An 'arrangement' was achieved, and with it the placing of contracts by the Russians in which Vickers duly had a manufacturing role, though on a much lesser scale than had been hoped. The more distant outcome was the formation in 1960 of a new company, Hans J. Zimmer AG., with headquarters in Frankfurt, in which Vickers had a holding of 50.5 per cent, followed in 1965 by outright acquisition by Vickers when Hans Zimmer died. In this way chemical engineering became a major—and, as it turned out, nearly disastrous—Vickers activity.

A second outcome of the Moscow visit was an order for two sugar-beet processing plants and the formation of a new company, Vickers and Bookers Limited, to carry out the work. Both plants were successfully completed but did not lead, as had been hoped, to any significant continuing activity for Vickers.

The Forward Development Department had also turned its attention to finding new interests for the Engineers and, perhaps a little oddly in the light of experience with tractor manufacture, put forward a proposition for the acquisition of All Wheel Drive Limited, a firm based at Camberley which initially specialised in the conversion of heavy vehicles to all-wheel drive but were now also engaged in the manufacture and sale, under licence, of Clark Michigan earth-moving equipment. The proposition was pursued and resulted in the purchase of a 60 per cent interest in All Wheel Drive in January 1961. Again, however, a project which at the time seemed full of promise, as a means of adding to the commercial work load of the Engineers, failed to fulfil expectations, and the Vickers holding was sold to Clark Michigan in 1964. The vehicle conversion activity remained with Vickers and was transferred to the works of the Vickers Onions subsidiary at Bilston.

With another venture it was a different matter. At its small Dartford works the engineering company had a lively business in the manufacture of steel shelving and filing cabinets, with the Government as its principal customer. In October 1960 the Company announced that it had bought Metchair Limited, a firm engaged at Hemel Hempstead in the manufacture of office and factory seating. This development, said the announcement, was 'in line with Vickers policy of extending its interests in the office equipment field'. A modest beginning, perhaps, but one which led Vickers by stages to a manufacturing agreement with Roneo Limited, to the acquisition of Roneo and to the formation of the Roneo Vickers Office Equipment Group with a bigger turnover, in due course, than either the Engineers or the Shipbuilders. The Engineers could reasonably complain that as soon as they had begun to develop a large and successful business it was hived away from them and made a business in its own right.

For the Shipbuilders, the Knollys years saw capacity well filled, a conjunction of notable achievements, modest to good profitability and the beginnings of a change of direction.

The inflow of orders reflected the ability of the yards to quote for work in most of the segments of shipbuilding in which requirement existed, whether naval or commercial. The major modernisation undertaken after the war had put the yards in very good shape and, though in some respects further investment had lagged, enough had been done to keep them technically in the forefront.

Orders for large liners were still available to be won, though the fleets of the big operators were now almost completed. Vickers had built many of the great post-war liners — *Himalaya* and *Chusan* for P & O, *Orcades*, *Oronsay*, and *Orsova* for Orient, and *Empress of England* for Canadian Pacific — and in 1960 Barrow was ready to deliver *Oriana*, at 42,000 tons the biggest liner ever built in England. At Naval Yard *Empress of Canada* for Canadian Pacific was nearly completed, with *Northern Star* to follow for Shaw Savill. Vickers also continued to get a full share of tanker orders, and in 1961 Barrow received the order from BP for *British Admiral*, the first 100,000-ton tanker to be built in Britain. In the same year Barrow undertook the building of a highly specialised tanker designed to carry liquid methane from the Sahara to Britain for conversion to gas.

Naval work continued to figure prominently in both yards, as it had always done, including Oberon class submarines at Barrow and frigates at Naval Yard. But in 1959 at Barrow Vickers had undertaken the building of Britain's first nuclear-propelled submarine — a development which was to transform British naval policy and cause Vickers to become almost exclusively a naval shipbuilder. For Barrow there was the proud day in October 1960 when the Queen, having launched *Dreadnought*, Britain's first nuclear submarine, went to an adjoining berth to inspect *Oriana*, the largest liner built in England. To round off that exciting year Barrow also delivered HMS *Hermes*, last of the Navy's conventional aircraft carriers. Since the reactor in *Dreadnought* was American she could not claim to be the first all-British nuclear submarine and this distinction went to HMS *Valiant* which Barrow built immediately afterwards. A little further ahead lay the Polaris submarines, armed with Polaris nuclear missiles as well as being propelled by nuclear energy.

In each of the six years 1956 to 1961 the shipbuilding company made a trading profit of between £1m and £2m, though Barrow had appreciably better figures than Naval Yard. The best year was 1959 with a profit of £1.7m.

Strongly linked with the shipbuilders, though a separate company, Palmers Hebburn on the Tyne kept itself busy and profitable with ship repairs during these years. The prospects looked sufficiently promising for Vickers to decide to equip Palmers with a very large dry dock. The scheme eventually cost £4.5m, and involved buying an adjoining site of some 18½ acres to give a total river frontage of nearly 2,000 feet. The new dock itself was 850 feet long, which was big enough to accommodate the largest tankers of the day, and an extension to 1,000 feet could be made without undue difficulty if 'giants' of 100,000 tons came into service. Unfortunately, the 'giants', when they appeared in the aftermath of Suez, had to be designed to go round the Cape, given the continued closure of the Canal, and so far exceeded 100,000 tons. This was a development that only a very exceptional crystal ball could have foreseen, but it meant that the new dock was destined to disappoint the hopes that had inspired it.

In the Canadian and Australian companies, as in the UK, more profit came from engineering than shipbuilding, though the achievements of the shipbuilders were more likely to catch public attention.

In 1957, the first full year following its re-acquisition, Canadian Vickers showed a trading profit of £1.2m, which must have been immensely encouraging to those who had advocated the return to Canada, especially Lord Weeks. It was a result which flattered to deceive, however, for 1958 and 1959 saw a recession in Canadian industry which reduced the figure, first to £700,000 and then to only a fraction above break-even. Fortunately, the tide turned and by 1961 the company was back to the million pound mark.

If financial returns were erratic, there could be no doubt that Canadian Vickers held a position of importance in Canadian industry, particularly in Quebec Province. At Montreal the company was strategically located on the St Lawrence between the sea and the Great Lakes. Very large vessels were not, therefore, part of its shipbuilding stock-in-trade, but its programme included destroyer escorts for the Royal Canadian Navy, icebreakers, fisheries patrol vessels, and dry-cargo bulk carriers to operate along the St Lawrence but also with ocean-going capability. Despite the relative smallness of its berths (including two totally enclosed), the Marine Division built a 26,000-ton bulk carrier in 1959-60 by the ingenious expedient of building and launching the fore and after ends separately. The two halves were then towed 170 miles westwards along the St Lawrence to Quebec where they were joined together in the large graving dock at Lauzon. This achievement was repeated, and as a variation on the theme, half of one ship was built at Montreal and the other half at Quebec by Geo. T.

Davie & Sons, a Canadian Vickers subsidiary. For bread-and-butter profitability, however, the Marine Division looked to shiprepairing rather than shipbuilding, and the repair facilities in both Montreal and Quebec were usually busy. In 1959, for example, repairs were carried out on 238 ships.

In the Engineering Division capacity existed to produce massive items of equipment for virtually all the main Canadian industries, from calandrias and lock gates to kilns and crushers. Following agreement between the Canadian and US Governments on a Defence Sharing Programme in 1959, the Division received substantial orders for missile tubes and hull sections for the American nuclear submarines, and this work was to provide a continuing base load of considerable importance to the company in the years ahead.

The activities of the Vickers companies in Australia had strong similarities to those in Canada, in the sense that the shipbuilding and shiprepairing operations had a strong naval content, while the engineering works at Melbourne and Perth had an important place in the national economy as manufacturers of capital equipment for the main industries, especially the mining industry. As in Canada, therefore, the fortunes of the companies were closely linked to the state of the economy and were likely to follow the same fluctuations. During the years from 1956 to 1961 their financial performance was far from impressive, and the new acquisition in Western Australia, Vickers Hoskins, had trading losses in three successive years. An important factor in these losses, however, was the disruption to business caused by removal of its activities from Perth itself to an industrial estate on the outskirts at Bassendean.

Very low profitability came also from the Cockatoo company in Sydney, though here the problem lay essentially in the nature of the agreement under which Vickers held a lease from the Commonwealth Government for operation of the dockyard and other facilities on Cockatoo Island, in the inner recesses of Sydney Harbour.

With poor results from Hoskins and Cockatoo it was from Ruwolts in Melbourne that the main profit contribution had to come. Between 1957 and 1961 the Ruwolt profit before tax varied from £126,000 to £382,000. If, overall, the Australian figures during the period were disappointing, they did not diminish the belief of Lord Knollys and his colleagues in London that in the longer run the energy and resources of this relatively young nation must ensure its prosperity, and that the Vickers investment would prove a good one. In token of that confidence they had approved not only the Hoskins acquisition, but also extensive capital expenditure by Ruwolts. They also examined a number of other propositions for further investment in Australia, but for various reasons these did not prove attractive.

In both Canada and Australia the 'rein' from London was exercised with a very light hand. The Boards contained a strong infusion of 'local' non-executives of proven ability and distinction in other fields, and one of them was usually chairman. Thus in 1959 the chairman of Canadian Vickers was Major-General A. E. Walford, a man of considerable and varied achievement, who held office until 1967. Though not a professional soldier he served throughout both world wars. During the first war he saw action at Ypres, the Somme, Arras and Vimy in ranks up to sergeant, and won the Military Medal before being commissioned. During the second war he served overseas for five years, returning to Canada in 1944 to become Adjutant General of the Canadian forces and a member of the Army Council. His services were recognised by a CBE in 1944 and a CB in 1946. Between the wars he first helped to set up an aerial services company, operating 18 flying boats, including a Vickers Viking amphibian, but then pursued a career in finance and accountancy. To this career he returned in 1946, though for seven years he found time also to be Honorary ADC to the Governor-General of Canada, Lord Alexander. His principal appointments were with the Morgan Group in Montreal, but he took other directorships also. He figured prominently in public life, both in Montreal and at the national level, and gave strong support to movements such as the Boy Scouts, the Salvation Army and the YMCA.

In Australia the non-executive chairman of Vickers Australia was Mr A. R. L. Wiltshire who joined the Board when the company was incorporated in May 1956 and remained in office until his retirement in 1964. He had been chief executive of the Australia & New Zealand Bank in Australia and was also a director of many other major Australian companies, ranging from mining and manufacturing to assurance and pension funds. Public recognition of his achievements in these and other fields, notably as a soldier, was evident in the awards of CMG, DSO and MC. To prove the rule by an exception, he was succeeded in 1964 by Captain G. I. D. Hutcheson, who came from inside the Group as Managing Director at Cockatoo. 'Hutch', an ex-naval engineering officer, was a man of powerful and likeable personality whose voice commanded respect both within the Vickers companies and in wider circles in Australia. A strong interest was taken by Lord Knollys in the Canadian and Australian operations, and he and Sir Charles Dunphie were members of both Boards, though attending only selected meetings.

Canadian Vickers had two chief executives during the 1957–61

4

period, first J. E. Richardson, OBE, who had been Deputy General Manager at Barrow Engineering Works, and then R. C. Pearse, an English-born Canadian, who had served an engineering apprenticeship with Stothert & Pitt before emigrating. He made a successful career in Canada, and joined Vickers from the Dominion Bridge Company Limited, holding the fort until 1967. In Australia the position was different in that Vickers Australia functioned as a holding company, of which Captain Hutcheson was the chief executive, with each of the main companies – Ruwolts, Hoskins and Cockatoo – having its own board and chief executive.

In India Vickers undertook a third 'imperial' venture, but this time as one of three partners, the others being Babcock & Wilcox and an Indian company, Associated Cement Companies of Bombay. The three came together to create a new heavy engineering works at Durgapur, near Calcutta, with an output mainly comprising cement plants, power station boilers and pressure vessels. Ownership was so arranged that ACC had a holding (49.8 per cent) exactly equal to the combined holdings of its British partners (each with 24.9 per cent), the very small balance being held by other Indian interests. The book value of Vickers holding was £747,000, to which was added mortgage debenture stock to the value of £378,750, to make a total investment of £1,125,750. The new works was inaugurated by the Chief Minister of Bengal early in 1962. Lord Knollys attended the ceremony and spoke in praise both of the works and the housing complex created for the employees. 'It is an admirable example,' he said, 'of British and Indian co-operation in the industrial development of India.' In fact, little benefit was to accrue to Vickers. Though competent as manufacturers, AVB had to struggle to make financial ends meet. In any event currency and other restrictions made it unlikely that any dividends could be paid across frontiers.

In India Vickers also collaborated with the Indian Government in two major defence projects, one involving the setting up of a factory to manufacture Vickers Battle Tanks and the other to build Leander class frigates for the Indian Navy. The agreement on tanks in 1961 involved the supply of tanks from Newcastle while the factory was being set up in India to undertake local manufacture. The delivery of tanks from Elswick began in 1965, and the first Indian-built Vickers tanks, re-named the Vijayanta, came from the production line in 1969. Total estimated Vijayanta production was put at over 500. The frigate project, in which Yarrow (Shipbuilders) Limited were also involved, covered expansion and modernisation of Mazagon Dock, near Bombay, together with the provision of technical aid and the supply of materials in building the frigates. The first frigate was launched by the Prime Minister of India, Mrs Indira Gandhi, in October 1968. By its nature the project was likely to be the last of its kind since the purpose of the Indian Government was to become self-sufficient in armaments. In a message to mark the launching, the Vice-President stated the aim clearly and colourfully. It was, he said, 'to steer clear of the shoals and eddies of foreign dependence and reach the shores of safety and self-sufficiency'.

Looking across the whole spectrum of Vickers affairs as he neared retirement from the chairmanship, Lord Knollys must have been very conscious of many hopes frustrated, but aware also that the Company had absorbed remarkably well the shock of having its aircraft business transformed into an investment in BAC. Powers-Samas also had lost its status as a wholly-owned subsidiary and become an investment in ICT. The fears about a further heavy diminution in armament work had not been realised, and with the Shipbuilders at the centre of the nuclear submarine programme, and the Engineers having a good expectation of sharing in the re-equipment of the Army with a new generation of tanks, capacity at Barrow and Elswick seemed likely to be kept busy. The running sore of the tractor project had been cauterised. First steps had been taken to move into chemical engineering and some other new areas. English Steel were in sight of opening the £26m. Tinsley Park plant. The Company's research effort had been stepped up through creation of a group research company. A group scheme had been established to raise the quality of management through new procedures for recruitment and career development. The institution of annual management conferences brought together all the most senior executives in the Group so that there could be a full understanding of the problems the Company faced and action agreed on how they should be handled. In particular, the need to protect liquidity had been driven home

Then, of course, there was the Tower. This was very much a Knollys project. It began formally with a minute dated 5th October 1956, from 'K' to General Dunphie. In view, he said, of the pressure on space at the Broadway headquarters, the inconvenience of the layout and facilities and the fact that the lease ran only for another 18 years, 'we ought to consider the future'. The prime objective should be for the Company to 'have our own Vickers House, preferably on a long lease, capable of holding all the London activities of the Group, with a reasonable surplus which can be conveniently let on short term'. The minute concluded by noting 'the proposition from Mowlems about the site between Millbank Hospital and Thames House'.

Whatever other inquiries might produce it was clearly this site which offered most attractions to Knollys. It covered an area of 3½ acres in a prime position on the bank of the Thames. It had once been the site of 'the dreaded Westminster Penitentiary', from which manacled prisoners were taken by barge to Gravesend for shipment to Australia. Now there was a row of twenty houses – thought to be about 100 years old – on the river frontage, and nine mews houses 'behind the Speaker's Stables', the rest being mainly used as builders' yards and depots by Mowlems. The freehold was owned by the Crown, with the lease held by Mowlems but due to expire at the end of 1958. The London County Council had declared the site to be suitable for a high building, and Ronald Ward & Partners, architects and engineers, had been commissioned by Mowlems to prepare plans. They envisaged 'using modern materials and advanced techniques, which would produce fine modern elevations in good taste, rather than attempting to compromise with past architectural styles which developed from quite different building techniques'.

It took little discussion to establish the central role of Vickers in the development of the site, with Mowlems as the contractors, and in March 1957 Knollys set up a small informal committee of directors to 'study and advise' on the development. This committee included David Pollock who was also a director of Legal and General. For him (and some others) it was questionable whether, at a time of heavy cash requirements, the Company should tie up money in a building project. He knew, however, that Legal and General were interested in property investment, and logic pointed to the possibility of an arrangement between Vickers and Legal and General which would achieve the ends of both. Again, little discussion was necessary to clear the way, and in April 1957 Knollys was able to write to the Crown Estate Commissioners to say that, though Legal and General would eventually become Head Lessees, they had authorised Vickers to negotiate the rent and terms of a lease. These negotiations led to the granting of a lease running to 2058, with Legal and General as the head lessees and Vickers as sub-lessees. In short, Legal and General agreed to finance the development and to lease it to Vickers for 95 years.

The architects' plans had meanwhile gone ahead. They provided for a tower block 387 feet high, at that time higher than any other London building, together with an 8-storey subsidiary block (which became known as the Y block because of its shape) and a 12-storey block of flats with a frontage in John Islip Street (a requirement of the planning permit). Though the height of the tower was explicitly stated, the number of floors this represented was arguable since the combination of a viewing gallery and housing for functional plant of various kinds at the top could be variously expressed as two to four storeys. Officially the figure was 34, but for practical purposes the tower block had 30 storeys for office use. It had two very striking architectural features which together set it apart from the design of other high buildings in London, all too often no more than variations on the matchbox theme. First was the use of curved convex and concave faces—a shape that might result if a conventional matchbox model had been picked up and squeezed between finger and thumb (leading to the possibly apocryphal story that this was precisely how the design was achieved). The second distinctive feature was the use of glass cladding, framed in stainless steel, so that, to quote the architects, one had the effect of 'the glass-clad concave and convex facades being mirrors with lively patterns of reflections'.

With the project firmly under way that might have seemed to be that, but Vickers now had an attack of cold feet. Was the scheme over-ambitious? Would it be economical? Did Vickers need a headquarters as splendid as this? Was property management really their business? Would the City and the stockholders consider it a wrong priority?

These doubts were reinforced when, at the beginning of 1959, Powers-Samas merged with British Tabulators. It had been envisaged that Powers-Samas, as a wholly-owned subsidiary, would be one of the principal occupants of the tower block. Perhaps, it was suggested, the new enlarged company, ICT, would be prepared to occupy the whole of the tower block, in which case Vickers would be content with occupation of the Y block. In fact ICT preferred to set up head-quarters in a building of its own at Putney. Vickers then pursued the idea of making the same offer to other major companies, and found that Courtaulds were interested, to the extent indeed that the Board approved a proposal that Courtaulds should be invited to take over the whole of Vickers interest, provided Vickers could have a lease for use of the Y block. Courtaulds interest also faded, however, and so the Board had to think again about its cold feet. To the informal committee of directors, Lord Knollys said that from the purely economic point of view the right course would be to occupy the Y block only, but this might not be acceptable from a prestige point of view. 'It is a case of balancing economics and prestige.' In fact, when the sums were done, it appeared that, if the Company occupied the Y block, income from lettings would be £478,000: if they occupied the top 15 floors of the tower it would be £450,000. With such a relatively small difference the Board opted for prestige.

Nor did the finding of tenants prove unduly difficult, and a large part of the surplus accommodation in the tower was taken up by the Ministry of Works for the housing of government departments, while the Electricity Council occupied the Y block. Several major companies also took floors in the tower then and later, including Pearsons, Plessey and Ferranti. Vickers did not wish to have responsibility for managing the block of residential flats, and Legal and General agreed to take this over, with one floor let to Vickers at a peppercorn rent as a hotel suite for staff visiting London from outlying establishments.

In the end, therefore, the scheme came to fruition in much the way that Knollys had envisaged. It was not, however, completed in time for him to be the first occupant of the Chairman's office on the 29th floor, though as Chairman of English Steel Corporation he was to have an office on the 18th floor. On the principle that, in a gentleman's residence, the best rooms face south, the Chairman's Office had been located on the south-east corner. In fact, this gave him Battersea Power Station as the centrepiece of his view. Had the north-east corner been selected he would have looked down on the Houses of Parliament and Lambeth Palace, with a magnificent panorama of Westminster, the West End, and the City, stretching to Hampstead heights. Immediately above the Chairman's office, on the 30th floor, was his private dining suite, small but elegant and embellished with six Piper montages. Other dining and committee rooms were also located on the 30th floor, and the Board Room occupied much of the north face. For the Board Room Misha Black designed an unusual and distinctive feature. Imposed on the curved inner wall, panelled in walnut, were silhouette carvings of famous ships built by Vickers since 1883. The panelling and carvings were all the work of craftsmen at Naval Yard and Barrow.

Though the formal opening ceremony did not take place until May 1963, when it was performed by Lady Knollys, Lord Knollys at least had the satisfaction of pouring out the last skip of cement at the topping-out ceremony in July 1961. In the open air, at nearly 400 feet above ground level, this was no occasion for those disposed to vertigo, but all went well, the weather was kind, and no calls were necessary for the sets of waterproof clothing held in readiness for the guests.

In assessing the six years of the Knollys regime statistically, various permutations and conclusions are possible, but on the basic measurement of dividend rate the period was static with 10 per cent paid in each of the six years. Dividend cover averaged just under twice for the period, with the last year, 1961, as the highest at 2.4 (in 1956 it had been 2.3). The 1961 figure was uplifted by an exceptionally low charge for taxation, however, because of an advantageous ruling on investment allowances in relation to prototype aircraft. Capital employed increased from £104m in 1956 to £147m in 1961, including an increase in share capital from £37.6m to £49.4m. With pre-tax profits at a lower level in 1959, 1960 and 1961 compared with the three earlier years (1956 produced £12.7m and 1961 £9.4m), the return on the increased capital employed (at book value) showed a marked deterioration – from 11.9 per cent in 1956 to 6.9 per cent in 1961 on a steadily downward trend. The stock market's reaction to the Company's results, and to events such as the hiving off of Powers-Samas and the formation of BAC, was reflected in prices fluctuating during the period between a high of 46/- in 1957 and a low of 23/6d in 1962. In general, Vickers was regarded as a good jobbing share, worth buying when its price fell.

The loss of the aircraft company and of Powers-Samas as wholly-owned subsidiaries had various statistical effects, notably an increase in the value of investments from £2.5m in 1956 to £26.3m in 1961 and a decrease in numbers employed from 90,930 in 1956 to 58,531 in 1961 (though the bill for wages and salaries did not decrease in anything like the same proportion, falling to £47.6m in 1961 from £54.2m in 1956). The total paid in wages and salaries during the six years was £317.5m. Taxation for the period totalled £31.2m and dividend payments £16m, with retained profits also totalling £16m. Reserves and retained profits increased from £35.8m in 1956 to £56m in 1961.

The pattern of sales showed an average of over £150m annually throughout the period, with a peak of £202m in 1957, when the value of aircraft sales stood at £70m, reflecting in particular the high success of the Viscount. Other factors apart, this figure was bound to decline progressively following the formation of BAC in 1960 since sales attributed to Vickers then came only from 'old account' aircraft. Shipbuilding sales fluctuated between £24m and £47m, depending largely on deliveries of major vessels. The Engineers had an annual average of £48.5m, without excessive peaks or troughs, and English Steel Corporation had a very consistent sales performance between £33m and £38m annually.

In his valedictory address at the Company's annual general meeting in June 1962, Lord Knollys said that he had been Chairman 'during a stimulating though not too easy period'. He referred in particular to the two main problems with which Vickers had been faced – how to fill the gap in armament sales, which had fallen by £20m between 1957 and 1961, and 'how to make it possible for a British aircraft industry to survive and to retain our own leading position in conditions financially practicable for any private manufacturer'. To the second problem, he said, the solution had been the establishment of BAC. As to the first, he described the action taken to obtain and develop commercial products, but recognised that the benefits had been slow to work through into profits. He thought, however, that they represented 'solid promise for the future'. Other topics singled out for mention were investment overseas (responding to the growing insistence in so many countries on local manufacture); modernisation and extensions in shipbuilding, engineering and steel; reorganisation and expansion of the research effort; streamlining of Group organisation; creation of a Forward Planning Department; and development of 'our methods of recruitment, education, training and the roads to promotion'.

The first objectives, he concluded, had been to 'fill the gap' and develop for the future. He believed both had been 'largely achieved'.

2

Birth of BAC

Engineers are a notably under-rated section of the community, or perhaps it is simply that their skills and achievements are taken for granted. Mostly they labour anonymously. They tend to live apart, preoccupied with what they are doing and not given to making a fuss about it. For the outside world it is difficult to comprehend the natural laws and principles with which the engineer grapples, or to wax lyrical about, say, a power press or a bottling machine, whatever its usefulness. With some of his products, however, the engineer compels attention without need for words or explanation. The beauty and power of great bridges and great dams are totally explicit, and most of us are fascinated by those products of engineering that carry us from place to place, the railway engines, the ships and the aircraft. If engineers as a whole are a race apart, the shipbuilders and the aircraft manufacturers have an inner world of their own. The danger for them all is that design and production becomes an end in itself, with profits and the market place playing second fiddle.

For the maker of aircraft the need to match technological skill with business and political acumen is vital. He also needs luck. To make and sell aircraft successfully involves fine judgments on a range of imponderables. It begins with the best calculation that can be made of requirements that will exist in ten years time. If that basic prognosis is wrong the project will collapse. Not only will it collapse but it will have wasted a great deal of time and very large sums of money. Even if the calculation is well made, extraneous factors may still come into play that will radically change the situation for the worse at the last moment. In the wake of an economic slump airlines will seek to make do with their existing aircraft for a few more years. Environmental pressure groups may delay critically the introduction of a new aircraft, especially if they are being manipulated for political or economic purposes. At the best, given a decade or more of design and development, continual modifications and refinements will either suggest themselves or be demanded by prospective customers. Costs multiply accordingly, and have to be borne by the manufacturer, who must therefore be prepared to put up large amounts of risk capital. Moreover, as costs multiply so does the number of aircraft he must eventually sell to recover development and production expenditure. So heavy is the financial burden that there is little prospect of carrying it unless the manufacturer has a base load of military work, and this in turn will fluctuate with defence policy and the timing of new experiments.

For the British manufacturer there remains the formidable hurdle of competition with his American rivals, nourished by massive military programmes and a domestic market vastly larger than the British and zealously guarded. Even if British airlines bought only British aircraft —and this is not their practice—the British manufacturer could not make ends meet on domestic orders alone. For sales on the scale he must have for profitability he has to sell to foreign airlines, and the foreign airlines must if possible include one or more in North America itself. An aircraft built to specifications by a British operator is bound to make losses, not only because orders will be too small, but also because significantly different specifications would almost certainly be required by foreign operators.

In the perspective of time, all these factors are readily identifiable, but in some instances they had not become apparent in the mid-fifties. Lessons had still to be learnt. One proved sharply painful. In designing the generation of aircraft to follow the Viscount, Vickers made its starting point the expected requirements of the national airlines. The Vanguard turbo prop resulted from a specification provided by BEA. The VC10 jet developed from a specification by BOAC. Both performed profitably in service, but for Vickers they were financially disastrous. Early development troubles with the Rolls-Royce Tyne engine seriously delayed production of the Vanguard, and when it eventually settled down it had ceased to be a marketable product, not least because Lockheed had by then rushed the Electra into production. The VC10 was technically a splendid aircraft, but it had been tied too closely to BOAC requirements and its fate was sealed when BOAC ran into a period of financial stringency (though, Vickers would argue, that was no reason for denigrating the aircraft in order to justify cancellation of orders).

The financial incubus of the Vanguard and the VC10 was to plague Vickers for over a decade. Though other major factors entered into the thinking which led to the formation of BAC in 1960, the need to obtain Government support for development of the VC10 decisively influenced Vickers' own attitude. To an extent the merger brought the desired backing, but 'old account' aircraft remained with the parent companies and Vickers had to continue to take financial responsibility for the Vanguard and VC10, with BAC as its agent for manufacture and sales. In the eight years from 1956 to 1963 Vickers' loss on the Vanguard totalled £16.7m. By 1963 it was also clear that heavy losses must be expected on the VC10 and a provision of £15.3m was included in the 1964 Accounts. This drain on resources throughout the period caused increased borrowing and severely hampered efforts to expand and diversify in other activities.

In 1956, however, when Lord Knollys became Chairman, all seemed reasonably well with the aircraft company. The Viscount was at the peak of demand and production, and on the military side Valiants were being delivered in large numbers to the RAF, while the Swift was being followed by the Scimitar for the Navy, the first sweptwing aircraft to be produced for the Fleet Air Arm. Trading profit in 1956 reached £3.1m, as it had done in 1955, and results in the same order were achieved in 1957 and 1958. Even so the signs were increasingly there to be read. To be set against profits of £3.7m on Viscounts in 1958 were development costs on Vanguard of £2.5m. Moreover, profit on military aircraft had fallen from £3.6m in 1957 to £1m in 1958.

With so much depending on a base load of military work, the delivery of the last of the Valiants in August 1957, and completion of the Scimitar line in 1960, seemed ominous when seen in conjunction with the 1957 White Paper on Defence policy. This foresaw the V-bombers being supplemented (and presumably supplanted in due course) by ballistic missiles, and the replacement of fighter aircraft by ground-to-air guided missile systems. A position had indeed been reached in 1958 when for the first time since 1915 Vickers had no military orders in hand at Weybridge. Fortunately, gloom was relieved towards the end of 1958 when the Government indicated its requirement for a tactical-strike-reconnaissance aircraft, the TSR2, to replace the Canberra, while at the same time making it clear that the contract would be placed only on the basis of a collaborative project between two of the major aircraft companies. Vickers and English Electric had no great difficulty in agreeing to collaborate and their joint proposals were accepted, with Vickers receiving the main contract. To that extent immediate anxieties about military work were relieved. Vickers had also gone ahead with work on guided missiles, particularly a private venture on an anti-tank missile, the Vigilant, which had proved sufficiently successful in trials to warrant cautious hopes for orders.

If in 1959 there seemed less to worry about in relation to military aircraft, the overall results of the aircraft company in that year were depressing in the extreme. They had, in fact, moved into the red. Profit from the Viscount, military aircraft and other activities had been more than wiped out by Vanguard costs at £5m. Nor were the prospects encouraging for the years immediately ahead. Further heavy expenditure on Vanguard was probable, and a continuing flow of cash would be needed for the VC10 programme. 1959 was thus a year calling for crucial decisions about the future of the aircraft company – sadly enough since 1959 also marked the 50th anniversary of the first order placed with Vickers for a flying machine (a rigid airship for the Admiralty).

In his speech at the annual general meeting in June 1959 Lord Knollys hoisted alarm signals.

It was more than ever necessary, he said, for the Government to appreciate 'the great and disproportionate financial burden' borne by Vickers and other companies in privately financed ventures such as the Vanguard and the VC10. 'Without firm and early support by the Government, through RAF orders or otherwise, to back up the initiative of private enterprise in maintaining design and production facilities and in forward planning, this country is more than likely, sooner than some people might expect, to find itself without a real aircraft industry at all.'

He pointed to the massive Government support given to American aircraft manufacturers, whose civil turnover was seldom more than 25 per cent of their total. 'All we ask,' he said, 'is for the same kind of practical far-seeing support which the US Government affords to its own private enterprise aircraft industry.'

What was at stake, he concluded, was the possible loss to the country of 'a great industry with its international prestige, the loss of technological talent, the loss of employment and the loss of its large, all-important exports. In our own case we have, in the last four years, sold to overseas customers aircraft to the value of £100,000,000.'

During the following month, July, Lord Knollys had meetings with several Ministers to discuss the situation in the knowledge that within the Government it was felt that there should be a merger in some form between two or more of the principal companies. Vickers assumed that there was no prospect of obtaining aid for the Vanguard and therefore concentrated its efforts on aid for the VC10. Other factors in mind were that Vickers and English Electric were working together successfully on the TSR2 contract and that de Havillands had plans for a three-engined jet, the Trident.

An approach was made accordingly to de Havillands to explore the possibilities of an association on the basis of Government aid for both the VC10 and the VC11 and for the de Havilland jet. The response from de Havillands was at first sufficiently encouraging for Lord Knollys to report to the Minister of Supply (Mr Aubrey Jones) that the omens seemed propitious provided the Government offered launching aid. Furthermore, he said, Vickers would discuss with English Electric the possibility of their joining a merged company. This approach quickly established that a favourable attitude existed in English Electric, and by September the position had been reached when the Minister declared willingness to provide government support, given the formation of an integrated company by Vickers, de Havilland and English Electric.

Snags now began to appear. De Havillands were having second thoughts. The Trident and the VC11 were to a large extent competitive projects, and with the Trident development well advanced de Havillands could see the possibility that in a merged company the VC11 would be given priority. English Electric took the view that it would be better for themselves and Vickers to make a 50/50 arrangement and then jointly take over de Havillands: Vickers argued in favour of a tripartite arrangement and this was eventually accepted by English Electric, though with some reservations.

Towards the end of October the Chairmen of the three companies met Mr Duncan Sandys, the new Minister of Aviation, and handed him a statement setting out the extent of the agreement they had so far reached. The statement drew attention to the special difficulties existing for de Havillands, including the fact that, unlike the other two companies, their shareholders depended entirely on aviation interests. They also had misgivings about bringing into the merger the company within their group which produced guided weapons and Blue Streak. From the Minister came an assurance that Government aid would be forthcoming if the merger took place.

In November, however, when further discussions were held between the three companies, de Havillands took a discouraging stance, and a week later, declared their unwillingness to join a tripartite merger. Since the Minister did not regard a merger between Vickers and English Electric as sufficient in itself, the question had to be posed of whether Vickers and English Electric should jointly seek to take over de Havillands, as originally proposed by English Electric. The terms of an offer were accordingly being prepared when on the 14th December, in a telephone call, the Minister made known his view that an enforced take-over could not be supported. Instead he suggested that Lord Knollys should see Sir Reginald Verdon Smith, the Chairman of The Bristol Aeroplane Company.

Two days later it became known that Hawkers were about to make an offer for de Havillands on agreed terms. The bid could have been contested by Vickers and English Electric, but on a quick consultation it was thought better to pursue the possibility of bringing Bristols into a merger instead of de Havillands. Though the Bristol order book was not strong their participation, Vickers felt, would make for a more rounded group and would bring in valuable design capacity. Furthermore, Bristols were known to be in favour of participation. So quickly did action proceed that on the 18th December – only four days after the Minister's telephone call and one day after the public announcement of Hawkers bid for de Havilland – agreement was reached in principle between Vickers, English Electric and Bristols on a merged company with participation on a 40:40:20 basis.

The Ministry of Aviation added to the rush of events by publishing on the 22nd December a document with the title 'General Outline of the Government's Policy towards the Aircraft Industry and the Nature of the Government Assistance which may be expected'. This stated formally what had been previously said informally – that the aircraft industry should concentrate the bulk of its technical and financial resources into four strong groups, two making airframes and guided weapons and two making aero engines. It also recognised the necessity for the Government to contribute towards the cost of developing promising new types of civil aircraft subject to participation in earnings from sales.

Moving from the general to the particular, the document went on to say that if the Vickers English Electric-Bristol merger materialised the Government would contribute to the cost of developing the VC10, the Super VC10 and the VC11 (assuming this went into production). Towards the VC10 it would contribute £6.4m, and to the Super VC10 and the VC11 50 per cent of the development costs up to a maximum of £3.85m and £9.75m respectively. These amounts, if expended, would total £20m. The document also spelt out the basis on which the Government would share the margin of earnings from sales.

In general, the three potential partners felt that the Ministry's statement set the scene satisfactorily and at once pushed ahead with discussions on detailed arrangements for a merger. It was decided that the VC10 and the Super VC10 should be excluded from the merger, but that the VC11 and the TSR2 should be regarded as projects falling within the responsibilities of the new joint company. Bristols' helicopter interests would also be excluded. So smoothly did the discussions proceed that it was possible on the 12th January 1960 to make a public announcement about the intention to set up the company.

Summarising the situation for the Vickers Board Lord Knollys said:

'We are achieving our purpose of obtaining Government support; we are increasing our interest in guided weapons and small aircraft; we are achieving a satisfactory spreading of risk; and we are effecting all this without any new issue of shares. On the other hand, we are bringing into the merger more than the 40 per cent equity share representation which we shall acquire.'

The legal processes necessarily took some months to complete, and the merger agreement was eventually executed on the 10th June to enable British Aircraft Corporation to come into operative existence on the 18th. Its issued share capital totalled £20m, held by Vickers, English Electric and Bristol Aeroplane in the proportions of 40:40:20. The Agreement provided that if any of the partners subsequently wished to withdraw from participation it must offer its holding to the other two before attempting to sell it to a third party.

It had been agreed that the Board of BAC should have an independent chairman, and an invitation to take up the appointment was accepted by Lord Portal, formerly Chief of Air Staff. Members of the Board were nominated by the parent companies in proportion to their holdings, with Sir George Edwards becoming Executive Director, Aircraft, and Lord Caldecote, Executive Director, Guided Weapons.

With Vickers' annual general meeting also taking place in June 1960, Lord Knollys was able to make topical references to these developments. Just a year earlier he had spoken in sombre terms about the critical stage reached in the evolution of the aircraft industry, and had called on Government to support the industry. Now he could report on comprehensive action to deal with the situation.

On the withdrawal of Vickers from direct production of aircraft he spoke philosophically.

'Vickers has an outstanding record in aviation, extending continually over 50 years,' he said. 'The Gun-bus, the Vimy, Wellingtons, Spitfires, Valiants, Viscounts have become part of aviation history. This great tradition and those who have helped to create and maintain it have been carried into British Aircraft Corporation. . . . As a result of this merger our stake in aviation is maintained, but financial risks are spread and shared, and outstanding resources, human and technical, are combined to deal with the competitive situation of the future. That is an encouraging prospect.'

Possibly he should have added 'in the longer term' because BAC could clearly not be expected to produce much financial joy for its parents for several years at least.

Also, of course, Vickers still had the Vanguard and the VC10 on its hands, though at least some contribution could now be expected to

VC10 development costs. Against the heavy outflow on these two projects it could also set continuing profit from the Viscount, though with over 400 Viscounts already sold the profit was more likely to come from spares than from further sales.

In relation to Viscount sales, Vickers still had a highly worrying problem to resolve. This was the prospect of a major bad debt in the United States. In the mid-1950s Capital Airlines Incorporated had taken delivery of 60 Viscounts and was so pleased with them that it ordered another 15. Unhappily Capital ran into serious financial difficulties during 1958 and was unable either to continue payment on promissory notes covering the original 60 aircraft, or to accept the additional 15. There was thus the double worry for Vickers of trying to minimise a possible major loss and of finding a customer for the 15 aircraft in production for Capital.

Efforts to do so eventually succeeded, largely through the good offices of Howard Hughes, who had a financial stake in Northeast Airlines. Northeast agreed to take 10 of the aircraft on credit terms. These 10 were duly delivered and then, trouble piling on trouble, Northeast also fell victim to the financial blizzard. With prolonged default on payments, Vickers were advised by their lawyers to repossess the aircraft, and this they did. One of the Viscounts had been lost in a crash but the other nine were at Boston Airport. Under American law, disposal had to be by auction, and R. P. H. Yapp, now Director of Contracts, went to Boston to represent the Company. Three of the aircraft were sold in the auction and the remainder were bought in by Vickers at the reserve price. Of these, one was sold by Ron Yapp across the table at lunch to an operator who had withdrawn from the bidding because he disliked the auctioneer's wisecracks. The remaining five were flown from Boston to Florida and then sold in ones and twos.

In dealing with the problem of Capital's default, Vickers felt that it would be wise to co-operate in efforts to keep the airline in operation. These efforts continued for over two years with many twists and turns – 'We went through terrible traumas' – but an answer was eventually found in the merger of Capital into United Airlines in June 1961. Vickers received various securities and assets under the agreement, and sold these satisfactorily, so that at the end of the day the loss was much less than it might have been – a total of £2,057,000 on the promissory notes. After United Kingdom tax relief there was a net loss of £964,000 and this was charged to Contingencies Reserve.

The episode served to highlight yet another of the hazards facing an aircraft manufacturer – the need, all too often, to provide credit facilities of various kinds to its airline customers.

There could be no doubt, however, that by most standards of measurement the Viscount was a remarkably successful aircraft, not least in its export earnings. Its sales reached 440, of which 356 went to overseas customers, including 147 to North American operators. The first scheduled operation was by BEA in 1953 and the final delivery was made in 1964. Vickers eventual profit on the Viscount, including spares, totalled £17m. Spread over a long period, and seen against the overall investment and cash requirements of an aircraft manufacturer, this must be accounted no more than a modest return from an exceptionally successful project.

An essential factor in the Viscount's success was the Rolls-Royce Dart engine. In any aircraft project the efforts of the airframe designer and manufacturer and those of the engine designer and manufacturer must be made to harmonise and synchronise as closely as possible. This had happened with the Viscount. With the Vanguard the synchronisation was less than perfect. BEA required the Vanguard to be powered by a new and more powerful Rolls-Royce engine, the Tyne. To perfect an engine is a matter of identifying and removing snags over a period. Snags in the Tyne took longer to eliminate than had been expected and delayed the production schedule. Even in the earlier production aircraft some trouble remained and noise and vibration levels exceeded those considered acceptable.

The Vanguard was thus off to a bad start, but its problems were more deep-seated than this. The decision to build another turbo-prop aircraft had been based on an estimate of medium-range market requirements. With perfectly sound logic it had been argued that the greater economy in performance of the turbo-jet would more than overcome the marginally technical performance of the pure jet. What had not been foreseen was the greater passenger appeal of the pure jet. The market potential of the turbo-prop was much reduced, though it might still have offered good prospects for the Vanguard but for the delays in development which enabled Lockheed to jump in with the Electra. Nor, as it turned out, did the 'double bubble' design of the Vanguard involving segregation of passenger and freight functions, commend itself to operators as widely as had been calculated. BEA wanted it that way, and so in a different version did Trans-Canada, but no other operator placed an order. Deliveries to both airlines began at the end of 1960 and the last aircraft was delivered in 1964. Orders totalled only 43 — 23 for Trans-Canada and 20 for BEA — and the eventual loss to Vickers was estimated at £16.7m, which almost exactly equalled the profit on Viscount.

With the heavy costs of the VC10 also before them, Vickers had

5

little option but to conclude that development of civil airliners purely as private ventures no longer had commercial justification. The BAC merger therefore made both Company and national sense, even if it only helped to ease the Company's immediate problems. For half a decade Vickers' freedom of action continued to be hampered by the financial burdens of the VC10, with little return from the investment in BAC. In the 1970s, however, there was a harvest to reap.

3

The Nuclear Submarines

The word 'submarine' appeared only once in the 1957 White Paper (Command 124) outlining future defence policy. This single mention did not even refer to use of the submarine as a unit in the Royal Navy, but only to defence by surface vessels against enemy submarine attack. In the section on 'Sea Power' the intention was declared 'to base the main elements of the Royal Navy upon a small number of carrier groups, each composed on one aircraft carrier and a number of supporting ships'. Under the heading 'Research and Development' the White Paper did say, however, that 'increased emphasis' would be placed on 'the development of nuclear propulsion for maritime purposes, which has great civil as well as naval importance'. As to the nuclear deterrent, medium bombers of the V-class would continue to be the means of delivery, supplemented by ballistic rockets supplied initially by the United States.

This, then, was the 'broad framework' under which long-term defence planning was to proceed.

Events during the next six years were to gallop away in an entirely different direction. The nuclear-propelled submarine was to become identified as the main weapon of British naval armament, and the nuclear propelled submarine armed with ballistic missiles was to become the main means of delivering the nuclear deterrent.

As guidance for long-term planning the 1957 White Paper could hardly have been further removed from reality. This may be more a comment on the doubtful value of sweeping exercises in planning than on the short-sightedness of experts. Possibly, of course, the experts were persuaded against their better judgment to forecast as they did, or possibly for security reasons they did not feel it possible to reveal what was really in their minds. By 1957 the actual course of events must have been foreseen in some inner rooms for by then the Americans had nuclear propelled submarines in operation and had embarked on development of a system for delivering ballistic missiles from submarines. Britain, too, had already taken the first steps towards developing nuclear plant for submarine propulsion through a contract placed in 1956 with Vickers, Rolls-Royce and Foster Wheeler.

Hindsight is easy, of course, not least in giving a foreshortened view of events, and it is enough to note that taken at its face value the White Paper gave some very peculiar guidance to those, like Vickers, who also needed to plan ahead in allocation of manufacturing capacity for defence work. It was open to Vickers, however, to make their own guesses, and they were sufficiently on the inside of official thinking to be able to do this, provided they did not stand too much in awe of planning pronouncements in White Papers.

The Americans were the first to do any practical work on the development of nuclear plant for fitting into the confined space of a submarine. As early as 1939 a report had been submitted by the US Naval Research Laboratory on nuclear propulsion for submarines, and immediately after the war, when authority was given to build an atomic power reactor in Tennessee, five naval officers were invited to take part in the project. The overall purpose of the project was to provide the United States with 'the first practical application of power derived from the atom', and the invitation to the naval officers represented recognition of the possibilities of application to maritime use. Since most of the theoretical work on reactors had been done, the task was essentially one of engineering. The most senior of the five officers was Captain Hyman G. Rickover, who had been a submarine engineering officer and later head of the electrical section of the Bureau of Ships. Rickover had seen the pre-war paper by the Naval Research Laboratory and was fired by a determination to see that the first marine application should be in submarines. If it could be done, not only would the submarine become for the first time a true underwater vessel, without the necessity to surface frequently to re-charge its batteries, but there would be an entirely new dimension to naval strategy.

Had anyone but Rickover been selected for this duty the course of naval history might have been different—certainly there would have been a much slower evolution—for Rickover was an altogether remarkable man, fanatically single-minded, ruthless and arrogant in pursuing his ends, and ready to scatter protocol, procedure and red tape to the winds. Presumably these qualities had been fostered by the harshness of his background and upbringing, for his father had been a tailor in a Jewish village in Russian Poland. Rickover was born there in 1900 and stayed in Poland with his mother and sister when in 1904 his father emigrated to New York. In two years his father had earned enough to send for the family, and it was as a boy of six that Rickover had his first sight of the New World. At school in New York and Chicago he was a lone wolf, working assiduously, and it was through the influence of a family friend that he gained entry to the US Navy Academy. Here again he was a lone wolf, but his driving efficiency compelled recognition and he forced his way slowly upwards in both sea and shore jobs. His jobs afloat were mainly on surface vessels, but for a time he served on a small submarine as engineering officer.

Now, with his energies focused on the single objective of adapting nuclear energy to submarine propulsion, he drove forward with characteristic ruthlessness, cutting corners and making enemies, especially among the top brass, but carrying sufficient conviction in a sufficient number of high places, to move rapidly forward. A land-based plant was built in the Idaho desert to develop a plant suitable for installation in a submarine, while the Electric Boat Company (old friends of Vickers) designed a suitable hull in mock-up form at Groton. In 1951, the Electric Boat Company received a contract to build the world's first nuclear submarine, and in January 1955 *Nautilus* was 'under way on nuclear power' from Groton.

To have Rickover as a friend was of great importance to the British. Fortunately he had emerged from the war with friendly feelings for the Royal Navy and an admiration for Lord Mountbatten. As an engineer he also had a considerable regard for Rolls-Royce. Stories inevitably accumulate about men like Rickover. By the mid-fifties he had reached such eminence that he could put into practice his belief that business should be done only with the people right at the top. With Mountbatten and the Queen as his allies, it was said, he believed that no obstacles were insurmountable. He had a high regard for Lord Hives, chairman of Rolls-Royce, but Vickers did not at first loom very large in his consciousness. The story is told that Lord Hives and Lord Weeks (then Sir Ronald) went to see him on one occasion, and since his brief had named Weeks as no more than a lieutenant-general he paid attention only to Hives. During the war, however, Weeks had been high in the esteem of General Eisenhower, and the latter, now President, heard that Weeks was visiting the States. A secretary came into the Rickover meeting with a message from the White House. Could General Weeks please find time to have lunch with the President during his visit? Weeks was thus established and he and Rickover ended on very good terms, which certainly did Vickers no harm, though perhaps no less important in practical terms was the relationship that Vickers had established with Electric Boat in the early days of submarine building.

When he launched *Nautilus* President Truman admirably summarised what had been achieved and what the building of a nuclear submarine involved:

New metals had to be produced. Wholly new processes for refining and using these metals had to be invented, tested and put into production. All sorts of new machinery had to be designed and built to specifications more rigid than anything attempted by American industry before. The whole complicated mechanism required to make the atoms break apart had to be designed to fit into this vessel's hull. Safety devices had to be worked out to prevent the ship's crew from harmful radiation. Special controls had to be developed so that the speed and intensity of atomic fission can be regulated instantly by the flick of a switch. And all this intricate mechanism had to be rugged enough to withstand combat shock from depth charges and other attack.

These were the processes which the British Government, having seen the success of *Nautilus* and her successor *Seawolf*, formally set in train in 1956 when Vickers, Rolls-Royce and Foster Wheeler were asked to combine in the construction of prototype nuclear propulsion machinery for a submarine. For this purpose Vickers formed a new subsidiary, Vickers Nuclear Engineering Limited, 'to control policy within the Vickers Group of engineering developments in the atomic field'.

Though a good deal of knowledge existed in Britain about the maritime application of nuclear power, the main research and development effort by the Admiralty since the war had been devoted to submarines using concentrated hydrogen peroxide as their source of propulsion. This was a method under development by the Germans towards the end of the war and an experimental U-boat, the U1407, had been salvaged after being scuttled at Kiel. It was now re-named HMS *Meteorite*, and to make it seaworthy and to arrange for it to be towed to Barrow, the Admiralty sent to Kiel Peter Scott Maxwell, a submarine engineering officer who had behind him a distinguished war record, notably with *Tally-Ho* whose exploits in the Pacific became the subject of a book. The Admiralty also arranged a visit to Germany by Ernie Davies, head of the Internal Combustion Engine Drawing Office at Barrow Engineering Works, to advise on further development by Vickers of the HTP method of propulsion. After seeing *Meteorite* on her way Scott Maxwell remained in Germany for some months to retrieve a complete set of replacement parts, together with the prototype of a larger engine. He then went to Barrow to join the team at the Admiralty

Development Establishment, financed by the Admiralty and manned by Vickers under the direction of Dr G. H. Forsyth. The further development of HTP propulsion was carried out by Forsyth whose team also included six Germans who moved to Barrow with their families. Of Forsyth it was said: 'A brilliant engineer who made enemies because he was always right.' He was later to contribute significantly to the engineering of the nuclear submarines. On moving to Barrow Scott Maxwell joined Vickers, eventually becoming the Company's chief executive in Australia and a Vickers Limited Director.

Two HTP submarines were built by Vickers, *Explorer* and *Excalibur*, and *Explorer* began her trials in 1955, the year when the Americans had put *Nautilus* into commission. It was quickly apparent that though in short bursts the HTP submarine could move at a remarkable speed under water, it was still in essence tied to the surface, unlike the *Nautilus* which was not only a true submarine but could also sustain a high speed under water. Moreover the HTP submarine was small and noisy and HTP itself a highly volatile substance. Because of the fireballs ejected from the exhausts when the engines were started *Explorer* was quickly dubbed 'Exploder' by her crew. The conclusion was inescapable and the Admiralty now accepted that future submarine developments must be nuclear propulsion. Nevertheless both *Explorer* and *Excalibur* joined the fleet, and continued in service until 1966 and 1970 respectively.

With the decision taken to 'go nuclear' conflicting opinions emerged, both in the Admiralty and in Vickers, about how policy should be implemented. One school of thought argued that the best course, and certainly the quickest, would be to approach the Americans for sale of a complete nuclear plant for installation in a submarine which would otherwise be British designed and built, while work went ahead on design of a British reactor for installation in a second submarine. The opposing school took a stand on national prestige and said that it would be better to take a little longer in order to have a completely British nuclear submarine from the start.

Within Vickers the arguments ran into cross currents of rivalry between the Engineers and the Shipbuilders at Barrow about who should be in the lead. Traditionally the Engineers had undertaken responsibility for machinery design and installation and the Admiralty had placed with them the main contracts for producing the machinery and for building the land-based facilities at Dounreay, in northern Scotland, to test it. The view taken by the Engineers, crudely expressed, was that building a hull was a relatively simple business for which the Shipbuilders could have a sub-contract. The Shipbuilders on the other hand took the view, again crudely expressed, that building a submarine was an art understood only by themselves and all the Engineers had to do was supply the equipment to go inside it.

In the argument about whether to seek to buy an American plant the Shipbuilders, in the person of Leonard Redshaw, then second-in-command to George Houlden, strongly favoured pushing ahead in this way. Not only would it be quicker, they argued, but a good deal was likely to be learned, and a good many corners cut, if Vickers could work with Electric Boat, who had now accumulated considerable experience in nuclear submarine building. The Engineers, with the vested interest of their two contracts and confident about the progress that was being made, refused to be impressed by these arguments and pleaded that they should be allowed to get on with it.

At the Admiralty Lord Mountbatten, as First Sea Lord, found the 'buy American' argument more persuasive, supported as it was by powerful voices among his advisers. Authority was accordingly given in 1956 for talks to proceed with the Americans and these had begun when the Suez crisis effectively brought them to a halt for a year. On their resumption no great difficulty was experienced in reaching agreement on purchase of a reactor compartment, and under pressure from Rowland Baker,* a key figure in the Admiralty's technical hierarchy, Leonard Redshaw and others, it was further agreed that the experience gained by Electric Boat in building *Nautilus* and its successors should also be made available. Admiral Rickover led the American side in the negotiations and his admiration for the achievements of Rolls-Royce caused him to insist that the contract should be signed between Westinghouse, Electric Boat Corporation, as designers of the American submarine reactor, and Rolls-Royce. Coupled with their work on the British reactor, this put Rolls-Royce in a very strong position, and briefly they had heady thoughts of becoming submarine builders themselves. With a little persuasion, however, they accepted that the task would be better left to the experience and professionalism of Vickers. Admiral Rickover had also stressed the importance in a project of this kind of undivided control, and this helped to put Leonard Redshaw in the driving seat within Vickers, fittingly enough perhaps since in his qualities of single-mindedness and determination he was something of a Rickover himself.

In the outcome, no one had much reason to complain. If Vickers Shipbuilders had succeeded in pushing ahead without waiting for the British reactor, work on the reactor continued in the firm knowledge that it would be installed in the second submarine. The building of

*Knighted 1968.

Oriana, largest liner built in England, launched at Barrow in 1959 by HRH The Princess Alexandra.

"Concorde" viewed from beneath the wing of a VC10.

Hiram Maxim's aircraft 1894.

r George Edwards explains a point of detail in fuselage construction to His Imperial Majesty The Shah
Iran.

ickers Main Battle Tank in action during trials. Vickers were also the design parents of "Centurion"
d "Chieftain".

Roneo, vintage 1911.

Roneo Vickers, vintage 1978.

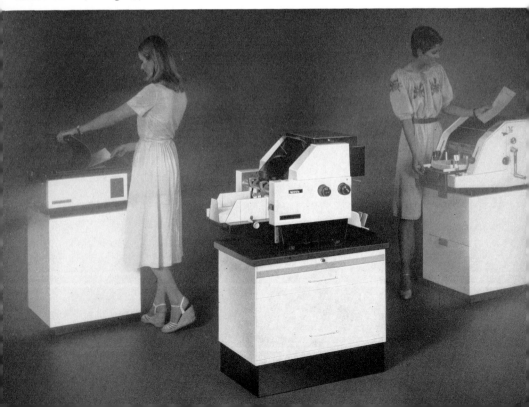

Dounreay continued without interruption so that Vickers Engineers retained a central role. The position of Foster Wheeler as designers of the pressure vessel and associated equipment was unaffected, and Rolls-Royce emerged in an even stronger position since they were now not only designers of the British reactor but also contractors for the American reactor. Above all, the decision to 'buy American' put a nuclear submarine into service with the Royal Navy three years earlier than would otherwise have been possible.

The building of *Dreadnought* at Barrow began in January 1959, her keel was laid by the Duke of Edinburgh six months later, she was launched by the Queen on Trafalgar day, 21st October 1960, and her delivery to the Royal Navy took place in April 1963. At every stage she was on schedule. This was a very remarkable achievement given the size and complexity of the vessel and the fact that she was the first in Britain to house a nuclear reactor. *Dreadnought* had the largest submarine hull ever built in Britain to that date. It was designed in tear-drop form on the basis of the high performance achieved with this type of hull by the US submarine *Skipjack*, the first to combine nuclear propulsion and the new hull design. In this again Britain had the good fortune to profit from American experience. The length of *Dreadnought* was 265 feet and her surface displacement was 3,500 tons. She carried a crew of 88 and her armament comprised six tubes for firing homing torpedoes.

In building this, their 295th submarine, Barrow had to use new methods of fabrication and welding, and in view of the urgency they went in for a good deal of ingenious improvisation. The hull was built in sections in the assembly shop. With each section weighing up to 150 tons, and ranging in shape from a cylinder to a cone, the task of moving them several hundred yards to the berth seemed formidable. It was tackled by use of the biggest road transporter in Britain. To house *Dreadnought* while she was fitted out after launching a new floating dock was necessary: it was built in HM Dockyard, Portsmouth, and towed to Barrow. A new drawing office had to be built to house the large teams of draughtsmen involved in preparation of the working drawings. A special progress department was set up, under the direction of G. G. Mott, and coloured pegs plotted on wall charts the daily progress of every item of equipment so that possible bottlenecks could be anticipated and removed. Outdoor welding of the special steel posed difficult problems, and to create the right conditions great shelters were erected over the hull. It was a case of learning all the time from experience, and the process drew strength from visits to and from Electric Boat. Vickers were fortunate also in their Admiralty contacts, especially Rowland Baker, who was to become part of the Barrow scene for a decade, first as Technical Chief Executive Dreadnought project, and then Technical Director Polaris Executive. A Vickers colleague said of him: 'A dominant character, a hard driver and a dedicated performer'.

With *Dreadnought* in its final stages, and the building just begun of *Valiant*, the first all-British nuclear submarine, a completely new and unexpected development took place early in 1963. The US Government had earlier agreed to make available to Britain the Skybolt missile which they were developing as a means of delivering the nuclear deterrent from aircraft. Simultaneously, however, development was proceeding of the Polaris missile for submarine firing. It was clear that at some stage a choice would have to be made between the two, or on a combination of the two, and in 1962 speculation was developing about the likelihood of the choice falling on Polaris. If missiles could be fired from under water the advantages were obvious, given the ability of nuclear propelled submarines to remain submerged for almost indefinite periods and to move about silently, so that the points of firing could be moved nearer to the targets without their location being detected. Authority to go ahead with the Polaris programme had been given in 1957, and in 1960, a Polaris missile was successfully fired from under water by the nuclear submarine *George Washington*. The Skybolt programme on the other hand was experiencing a good deal of trouble and the missile's reliability was still in doubt in 1962. In all these circumstances, though no final decision had been announced, opinion in the top echelons of US Government was moving towards the choice of Polaris. This would, of course, involve a major switch in US defence policies since responsibility for delivering the nuclear deterrent would be transferred from the Air Force to the Navy, with all that flowed from such a decision. For Britain, however, the switch would have an extra dimension. Though agreement existed for Britain to buy Skybolt no formal agreement had been made about Polaris, beyond a verbal assurance by President Eisenhower. If the Americans terminated Skybolt, and refused to replace it with Polaris, Britain would have only bomb-dropping aircraft as her means of delivering the deterrent. Against Russian possession of ballistic missiles, including almost certainly Polaris-type missiles, and constantly improving defence against aircraft attack, the British deterrent would then cease to be credible and she would be vulnerable to nuclear attack. Given this prospect it might be necessary to embark on a crash programme to develop a British missile for underwater firing – perfectly feasible but costly, and with a time gap of several years.

Coincidentally, President Kennedy and Prime Minister Macmillan were due in December 1962 to hold at Nassau, in the Bahamas, one of their periodical meetings to review international affairs in general and Anglo-American relations in particular. Usually these meetings held little drama but at Nassau, on the Skybolt-Polaris issue, there were all the elements of a confrontation since the State Department certainly, and possibly the President himself, felt that the proposed switch to Polaris provided an opportunity to edge Britain out of the nuclear power club. No doubt they would have preferred it not to happen under the aegis of a President–Prime Minister meeting, but the time scale of events inevitably put the issue on the agenda and, in any case, the Prime Minister was determined to establish the British position with absolute clarity and to leave the President in no doubt about the damage that would result to Anglo-American relations if the US Government terminated the Skybolt agreement without offering to make a similar agreement on Polaris. In the upshot, tough talking led to a compromise: the Polaris system would be made available, with British nuclear warheads fitted to the missiles, but the British Polaris submarines would be 'allocated' to NATO. Even so no missile firing could take place without the Prime Minister's authority and a right remained for the submarines to be withdrawn in specified circumstances.

The agreement in principle at Nassau was followed by immediate action to implement it. The British Government authorised the building of four Polaris submarines (operationally five would clearly have been preferable but cost considerations ruled out the fifth), design began in January 1963, a Polaris sales agreement was signed with the United States in April, and in May Vickers were appointed lead yard, with themselves building two of the submarines and Cammell Laird & Company Limited the other two. The contract was placed on the understanding that the work should proceed to a very strict time-table, with Vickers delivering the first submarine, HMS *Resolution*, on a schedule that would put her into service during 1968. To this end guarantees were sought, and obtained from the trade unions concerned. The Ministry of Defence (Navy) set up a Polaris Executive with Rear Admiral Rufus Mackenzie as Chief Polaris Executive. Leonard Redshaw, now managing director of Vickers Shipbuilding Group, was named Builders' Chief Polaris Executive.

To inject this volume of work into an already heavy programme at Barrow, including *Valiant*, *Warspite* and *Churchill* in succession to *Dreadnought* in the series of nuclear hunter-killer submarines, obviously made immense demands on the resources and organisational ability of the Yard. If in terms of size the attacker submarines had represented a

new concept, *Resolution* and her sister ships were to be twice as big —
in effect hunter-killer submarines with a missile section inserted into
the middle capable of carrying and firing 16 Polaris missiles. With a
length of over 140 yards, they had a tonnage of 7,500 tons surfaced and
8,400 tons dived, which put them into the same category of tonnage
as a cruiser. The crew numbered 141, compared with 103 in the later
attacker classes.

Some impression of the complexity of the builders' task can be
given by statistics. Over 10,000 detailed drawings were required and
nearly 1,000 fully illustrated manuals. Over 550 manufacturers contributed some 300,000 items of equipment, including highly sophisticated
mechanical and electronic equipment. Computers checked the progress
of each item to ensure that it arrived at the right place at the right time.
Strict quality control procedures were instituted. A full scale wooden
mock-up was built to ensure that snags were detected in the transition
from drawing board to three-dimensional reality (and also to serve as
a preliminary training ground for the crew). The hull was prefabricated
in 15 sections, the fore and aft sections then being assembled and
welded on the berth, with a space between them into which the
missile section could be fitted. As compared with American practice,
the missile tubes were built into the missile section during prefabrication
and not on the berth.

Much had already been learnt from experience with *Dreadnought*,
of course, and the work involved on the hunter-killer submarines had
caused a large increase in technical staff at Barrow. Even so numbers
had to be built up still further, and very rapidly, by the demands of
the Polaris programme. Not least, as lead yard, Barrow had to supply
Cammell Laird with specialised materials, training assistance in high
technology, and all drawings and test documents, and also ensure that
experience gained in building *Resolution* was fully available for the
building of *Renown* at Birkenhead.

Resolution, Barrow's 300th submarine, was on the slipway for only
30 months from keel laying to launch. The launching ceremony was
performed by the Queen Mother in September 1966, and commissioning took place in October 1967. By then the building of Barrow's
second Polaris submarine *Repulse*, was under way. *Repulse* was launched
in November 1967 and commissioned in October 1968. Vickers had
thus completed on schedule and to estimate her task of building two
Polaris submarines, while still working on the hunter-killer programme
and also completing the largest tanker built in Britain to that date,
British Admiral, and a new type of methane gas carrier for which she
was lead yard.

With the delivery of *Repulse* Barrow might have well have experienced difficulty in providing continuing work for the very large staff of technicians now assembled there, but orders continued for the hunter-killer fleet and in November 1968 came the contract for the first of a new type of destroyer for the Royal Navy. This was the Type 42, a guided-missile destroyer, to be launched in due course by the Queen as HMS *Sheffield*. Barrow also set up a department to meet the demand for highly sophisticated weapons, and this absorbed some 200 of the men who had been working on Polaris. Others were to find a niche in oceanics work which Leonard Redshaw rightly saw as an activity with great growth possibilities.

In 1969, the year after *Repulse* had gone into service, defence cuts led to a slowing down in the nuclear submarine programme. Until then Cammell Laird had shared in the work, but the Government now decided that contracts must be placed with one yard only. Their nomination was Vickers. The decision was strongly attacked by the Merseyside MPs and others who thought that it would have serious repercussions for employment in the area. The Government stood firm. Mr John Morris, Minister of Defence for Equipment, said in the House of Commons on the 4th March 1969: 'The simple fact is that our future nuclear submarine building programme can provide work for only one yard, and Vickers were chosen because of their clear lead in design capability and experience'. The following week in the debate on the naval estimates Dr David Owen, then Under-Secretary for the Navy, said in the House: 'We regard both firms as highly valued shipbuilders, but as a customer with all the relevant information available to us we have made a straight commercial decision,' (*Hansard* 10 March 1969). When the issue was put to the Prime Minister (Mr Wilson) in a Parliamentary Question he was even more forthright. 'The decision,' he said, 'was based on sound economic reasons and taken in the knowledge of Vickers' greater experience and capacity in all aspects of the work,' (*Hansard* 27 March 1969).

Questions were also asked in the House about the cost of Polaris and fleet submarines. The figures for *Resolution* and *Repulse* were given as £40.2m and £37.5m respectively, and for *Renown* and *Revenge* as £39.5m and £38.6m. For the three Fleet submarines completed to date *Dreadnought*, *Valiant* and *Warspite*—the cost was put at £18.5m, £24.9m and £21.5m respectively, (*Hansard*, Written Answer, 13 May 1970).

On contract prices in the circumstances of a single contractor the Minister of Defence for Equipment (Mr John Morris) told the House in a Written Answer (*Hansard* 26 February 1969) 'A fair and reasonable price will be agreed before each contract is placed on the basis of complete equality of information between the firm and the Ministry of Defence. The contract will include provisions to encourage the firm to achieve the maximum possible cost savings during construction and provide that the benefits of these savings should be equitably shared between the firm and the Ministry.'

In practice Vickers and the Ministry of Defence (Navy) made independent technical estimates and then went through these estimates together, section by section. This provided the groundwork for a commercial negotiation with the MOD Director of Contracts. The negotiation covered the overhead rate to be applied (with provision for annual review), escalation in wage and material costs, and a profit addition. The profit addition would be regarded as 'fair and reasonable' by the Ministry and as barely adequate by Vickers, having regard to plough-back needs. A loss could result if miscalculations had been made in the technical estimates, which was unlikely, or if delays had occurred through industrial disputes, which was less unlikely. In fact, few delays were experienced on *Dreadnought* and her immediate successors, and none of significance on the Polaris submarines. Later the record was to be less satisfactory.

Though the profits might be modest, it would not have occurred to Vickers to contemplate the possibility of not building for the Royal Navy. They had gone into shipbuilding towards the end of the last century as naval shipbuilders, 'to supply ships with their engines complete and equipped with guns and armour plate, entirely manufactured by the Company in its own works'. Ironclads were quickly followed by the first submarines for the Navy, and in each succeeding year the yard at Barrow was rarely less than half occupied with naval work. Great pride was taken in the work. It was seen less as a source of profit than as the performance of a national duty. This was an attitude in Vickers which drew criticism from financial purists, but in the long run it was perhaps not only a commendable sense of responsibility but also made sound commercial sense. Relations between Vickers and, first, the Admiralty and then the Ministry of Defence (Navy), were so closely interwoven over the decades that it would have seemed out of the question for either to seek to be less than 'fair and reasonable' with the other: indeed, the intimacy of knowledge was such that neither could have outsmarted the other even had they wished. Vickers built naval vessels with the highest degree of skill and competence. The Navy department knew this and valued it. They knew also that dealings with Vickers would be 'fair and reasonable'. As a result they turned to Vickers with most 'first of class' projects — as with the nuclear submarines and the Polaris submarines—and Vickers could count for most of the time on a hard core of naval work, a position of strength which was to stand them in good stead when the shipbuilding industry later ran into recession.

Part Three
1962–1970

Hard Going 1962–66

TOWARDS THE END OF THE KNOLLYS CHAIRMANSHIP VICKERS WAS already moving into a period of lower profitability and mounting problems.

At the heart of the matter was the cost of financing the VC10 which pushed borrowing to uncomfortable and costly levels. Even more damaging than the effect on profits of heavy interest charges was the inhibition placed on the Company's ability to provide funds for the development and expansion of its other activities. At a time when the Engineering Company in particular stood in need of new investment little or no finance could be provided. As but one example, plans to expand the small but successful brass business at Newcastle were frustrated by lack of cash for an acquisition costing less than £100,000. For the first six months of 1964, indeed, there had to be a moratorium on all capital expenditure.

The one consoling feature in the situation was that its end could be foreseen. A point had to be reached when the VC10 had exhausted its financial demands, when for better or for worse its future rested on the ability to make and sell rather than to develop and modify: but this point still lay two or three years ahead when Sir Charles Dunphie succeeded Lord Knollys as Chairman in June 1962.

It must have been very clear to Sir Charles that the scene was set for an uncomfortable start to his chairmanship. The enforced standstill in plans for growth was disagreeable enough, but other developments were also running against the Company.

The new works at Tinsley Park, on which £26m had been spent, was coming into operation at a time when the steel industry had moved into recession so that full advantage could not be taken of the new facilities. Nor was it simply a matter of frustrated promise at Tinsley Park: ESC as a whole had reduced profits. In the late 1950s ESC customarily returned profits of between £4m and £5m. In 1962 they fell to £2.8m and then to £2.2m in 1963: 1964 saw a small recovery to £2.6m and 1965 a return to £2.2m. These figures came at a wrong time in other respects. In 1963 the Macmillan Government was suffering setbacks which pointed to the possibility of the return of a Labour Government in the foreseeable future, and with it as a high priority the nationalisation of the steel industry. This possibility, and the uncertainty it engendered, became reality in October 1964, and so Vickers, having merged its aircraft business into BAC, now had to contemplate the enforced take-over of its steel business – the business in which it had been founded in 1827.

The falling away in results from steel formed only part of a disturbing pattern. Both the Engineers and the Shipbuilders were also feeling the impact of the recession. In the annual accounts Engineering and Shipbuilding appeared under a single heading so that for the outside world the precise performance of the two activities could not be identified. This form of presentation sprang less from a desire to blur the information than from the nuances of separating Shipbuilding and Engineering when at Barrow in particular, and to an extent at Elswick also, a good deal of engineering work was related to shipbuilding requirements. For internal purposes, however, a line was drawn, and the profits of the Engineers declined from £3.8m in 1962 to £2.5m in both 1963 and 1964, with Elswick and Crayford having poor years.

The Shipbuilders had an even gloomier tale to tell. For three successive years, 1964 to 1966, they were in the red. The trouble lay primarily in results from Naval Yard, at Newcastle, which showed a loss in five of the six years beginning in 1962. Despite its name, Naval Yard had more commercial than naval work and, like most British shipbuilding firms, was feeling the pinch of intense Japanese and other international competition at a time of deep recession. Contracts were taken at unattractive prices, and escalating costs could not be recovered in improved productivity. In particular, a provision of £3m had to be made in 1966 against a contract for five cargo liners for Alfred Holt & Company. Shiprepairing was also in the doldrums, and after 1961 the Palmers Hebburn business steadily returned small losses, with the exception of 1965 and 1968. Though Barrow Shipyard remained in profit (in 1963 by a margin of only £45,000), they could not initially reflect the appearance of nuclear submarines in their order book since profits could be taken only with deliveries (later, when the flow of production had become firmly established, it was agreed that profit could be taken in stages). Not until 1967 did the Shipbuilders return to overall profitability, but they then did so handsomely with a figure of just under £2m, following a loss of £1.4m in 1966. By then the work load at Barrow Shipyard was entirely naval, the 100,000-ton tanker for BP, *British Admiral*, having been delivered in 1965. The yard had also taken over from Barrow Engineering Works in 1964 the Marine Installation Department, a usefully profitable activity.

With the Canadian economy also at low ebb in the early sixties profits from Canadian Vickers slumped from £1m in 1960 to no more than break-even in 1963. The shipbuilding business in particular was in deep trouble. In comparison, trading profit by the Australian companies was at least remaining at a steady level, in the order of £0.5m, though the lack of growth was disappointing.

One seemingly encouraging development in 1963 was a profit of £0.7m from the Zimmer business, the first indication that chemical engineering might prove to be a successful new activity. Even here, however, some members of the Board expressed disquiet about the extent to which calls were being made from Germany for loan guarantees. It was pointed out, in reply, that the actual amount of Vickers investment in 1963 was less than £200,000, and that for a relatively small sum the Company had acquired an interest in the skill and knowledge of a man of exceptional inventive genius. Unhappily and unexpectedly Hans Zimmer died before the year was out, still a young man, and to the extent that the investment had been in one man it was lost. In fact, Vickers extended its interest in the business by purchase of the Zimmer family's holding and so became firmly committed to chemical engineering as a major new activity.

With so many activities in trouble during 1963 and 1964, and with bank and loan interest increasing sharply, the pre-tax profit for these years – the first two full years of the Dunphie chairmanship – showed a considerable decline: from £9.62m in 1962 to £8.46m in 1963 and £5.66m in 1964. Nevertheless the dividend was kept at 10 per cent, and in 1963 at least it met the Knollys criterion of being twice covered. Even in 1964 it was covered, though only just.

In 1964, however, it had been necessary to grasp the nettle of VC10 losses. For a long period there had been a running battle with BOAC about their wish to pull out of their commitment to buy the aircraft in the numbers originally agreed. No topic seemed to come before the Board more frequently and to be more vexatious. The VC10 had been developed to meet a BOAC specification, based on the need to operate in and out of airfields in Africa and South East Asia. On that basis 35 aircraft were ordered. With the need developing to operate it also on transatlantic services, a larger version was developed as the Super VC10, and BOAC ordered seven of this version to make their total order 42. The 'mix' was then changed to 28 standard and 12 Super VC10s – two less in total though without diminishing the total value of the order. The real rows began as Sir Giles Guthrie succeeded Sir Matthew Slattery as chairman of BOAC and sought, on economic grounds, to reduce the order to the smallest number possible so that there could be consolidation on a fleet of Boeing 707s. In justifying this policy he was believed by Vickers to have made damaging criticisms of the VC10, to the extent indeed that its prospects of sale to other airlines were seriously eroded. This Sir Giles denied, but the issue became one of national importance involving the Minister of Aviation. Vickers were affronted also by BOAC evidence given to the Select Committee on Nationalised Industries which seemed to imply that BOAC had been compelled against their wish to buy British, and that the VC10 had been designed for the Atlantic route. Sir George Edwards sought and obtained permission to appear before the Committee and established that the aircraft derived from a BOAC specification for tropical routes (high temperatures and short runways, coupled in some instances, with high altitudes) and that this specification was in no way put forward under 'Buy British' pressures.*

With the RAF also interested in the VC10 (though a diminished interest with the decision to use Polaris submarines for delivery of the nuclear deterrent instead of Skybolt), the possibility existed that a compromise position might be reached, with ministerial aid, as between BOAC and the RAF. In the end BOAC took 30 VC10s—18 standard and 12 Super—and in due course paid cancellation charges on 10 aircraft, totalling £7.5m, calculated on a formula designed to ensure that Vickers neither lost nor gained on the transaction. Other orders came from the RAF (14), East African Airways, Middle East Airlines and Ghana. In all, 54 VC10s went into service—to the operational satisfaction of the RAF and to the operational and commercial advantage of the airlines—but this total offered Vickers little prospect of recovery of its development cost.

With this position reasonably well established by 1964 Vickers had to provide in that year for the estimated under-recovery of development costs. The figure was put at £15.3m, which was carried to reserve accounts and did not appear in the Profit and Loss Account. The provision was designed, as it had to be, to meet the worst possible position, and in 1965 it was possible to write back £3m. Over the next 10 years further recoveries were made and the ultimate result was probably a break-even position. The fact remained that from 1960 to 1964 the financing of the VC10 seriously limited the Company's

*House of Commons Paper 9 June 1964: Report from the Select Committee on Nationalised Industries. Volume II. Paragraphs 1543-1573.

freedom of manoeuvre in its efforts to build up other activities, particularly in the Engineering Group.

During this period, however, simply in terms of the Profit and Loss Account, Vickers did receive a modest return from its aviation interests. If the return from the investment in BAC was small, there was a steady business in meeting the need for spares and replacements for 'Old Account' aircraft, including military aircraft. From this business in the three years 1962 to 1964 profits were £1.7m, £2.1m and £2.5m.

For W. D. Opher and his colleagues in the Engineering Group sorting out winners and losers was the priority task during these years, and if little could be done for the time being to strengthen the winners decisions could at least be taken about the losers.

Weakness in dealing with losers was not infrequently alleged against Vickers, and instances are not difficult to cite (at any rate, with the advantage of hindsight), but in the circumstances of the moment the issues rarely seemed clear cut. Reasons for weakness had first to be identified and a view then taken about the extent to which they were transitory and remedial. Were there management failings? Did a strong enough market exist into the future? Was the product good enough and if not could it be made good enough quickly? Was it available in a wide enough range? Could it be made more competitive by streamlining production methods, possibly in a merger of facilities with allied products? Having posed questions such as these, and hopefully arrived at sensible answers, it might still be concluded that pros and cons were finely balanced. Other considerations would then come into play. If in a given division there had been under-recovery of costs over an uncomfortably long period could an improvement in trading conditions be foreseen? If so, it might be worth while as an expedient to accept work at little or no profit. The alternative was to reduce costs which primarily meant dismissal of employees. Apart from an instinctive reluctance to do this, heavy redundancy payments would be incurred. With an upturn in trading conditions recruiting would have to start again, and the recruitment of skilled men was usually a slow and difficult process.

Many pressures therefore existed against closure or major retrenchment, and these pressures would usually reinforce a natural tendency in local management to seek to hang on as long as possible. Responsibility for final decisions rested with management at the centre, but even management at the centre might find it difficult to be fully objective and impersonal, especially in an ambience such as Vickers where attitudes were traditionally paternalistic and top managers usually men who had worked together for many years and were often friends as well as colleagues. In short, ruthlessness and bloody-mindedness were not sufficiently Vickers characteristics to ensure that necessary but harsh decisions were taken in good time.

The Engineers nevertheless went ahead with a good deal of action designed both to deal with losers and to encourage winners, provided the encouragement did not include heavy expenditure.

Much of the effort was focused on the southern works. Two works were closed – Thames and Weymouth. The Thames Works at Dartford was engaged on a small scale in producing explosives as a legacy from the armaments era. Its closure, said the official notice, resulted from 'the reduced requirement for conventional military explosives and the insufficient demand for commercial explosives'. Weymouth was also an armaments legacy as a specialist in the design and manufacture of torpedoes. With decreasing demand an attempt had been made in 1959 to rescue the works by transferring to it the Vickers hydraulics activities previously located at Newcastle and Wakefield. Two years later agreements were reached with the Racine company of Milwaukee for the rights to manufacture and sell each other's products. With both Vickers and Racine products being manufactured, capacity for the additional work was allocated at South Marston as well as at Weymouth. This was the situation when stream-lining of the Engineering Group began in order to cut out losers and to concentrate ongoing activities in a limited number of locations, with South Marston as one of these locations. The Weymouth Works was sold accordingly and its hydraulics activity moved to South Marston where a Hydraulics Division was set up. It had been hoped that a substantial number of the skilled personnel would move to Swindon, but in the event only 17 did, partly because jobs were available with the new owners of the Weymouth Works. In its new environment the hydraulics business, though now consolidated on one site, had the problem, among others, of training new operators, and these particular difficulties, added to the general difficulties being experienced by all manufacturers of hydraulic equipment, meant that South Marston had on its hands not a growth activity but an ailing one.

Nor were hydraulics South Marston's only problem child. The bright hopes placed in hovercraft development were fast evaporating. The hovercraft could attract headlines but not customers. The first vehicle built at South Marston, the VA-1, was essentially for research purposes. It became operational in 1960 and was used for testing, with speeds up to 40 knots over water. Though not large it was still too cumbrous to be taken overseas for demonstration, and VA-2 came into being as a vehicle small enough to be airfreighted to demonstration sites, but large enough to prove the practicability of the concept (in round figures it was 28 feet long, 15 feet wide and 10 feet high). VA-3 was considerably bigger (52 feet long) and was designed as a fast transport for 24 passengers in estuary conditions. It became the world's first hovercraft licensed to carry fare-paying passengers when an experimental ferry service was operated between Rhyl and Wallasey during the summer of 1962. Work on other variations also went ahead, including two hovertrucks, and later a hovertrain and a vehicle for carrying transformers and other heavy equipment for the Central Electricity Generating Board. For VA-4 a craft of about 100 tons was envisaged, capable of speeds between 70 and 80 knots carrying passengers and cars on a short-haul ferry service. Though licensing agreements were made with companies in the USA and Japan, it became increasingly apparent that the hovercraft would need time to establish itself as an acceptable form of transport, and that even then the market was likely to be limited. Perhaps the writing was on the wall when a potential customer from Denmark, contemplating services in the Kattegat, hurriedly backed away on being told that the passengers would need to wear seat belts, as in an aircraft: no ferry service could possibly be viable, he said, unless the passengers could stand at a bar. Reluctantly Vickers took the decision to incur no further development on either civil or military applications unless firm orders had been placed. For a time at South Marston a blind eye seems to have been turned on this decision, but eventually it was accepted that ways and means must be sought to put hovercraft design and development into a broader national setting. These efforts led in 1966 to an agreement with Westland Aircraft Limited to form British Hovercraft Corporation, in which Westland had the majority holding and the National Research Development Corporation also took part.

To a considerable extent South Marston had become a test bed for new products. If some, like hovercraft, proved disappointing high hopes were still held for others, and their promise was recognised by the formation of three new divisions – Automated Systems, Radiation Equipment ('to continue the development, design, manufacture and sale of linear accelerators and other devices using similar types of technology developed by Vickers Research Limited') and Food Engineering ('to be responsible for all aspects of the freeze-drying business'). South Marston also became in 1963 the manufacturing centre for high-pressure oxygen therapy equipment resulting from research at Sunninghill.

At Crayford, too, major changes were taking place, designed in this instance to release surplus capacity by concentrating on selected products, of which the most important were packaging and bottling machinery. This was essentially a retrenchment exercise and involved some 200 redundancies.

At the neighbouring Dartford Works, however, facilities were being extended following an agreement for Vickers Engineers to manufacture for Roneo certain types of steel office equipment, in addition to Vickers own production. It was a business in which Vickers was already well established, characteristically enough as the result of an effort to use wartime skills in peacetime production after the First World War. From hand-made wooden aircraft propellers the craftsmen in the Woodworking Division at Dartford did not find it too difficult to switch their skills to wooden office furniture. Subsequently, when Government offices began using steel furniture, Dartford responded by designing and manufacturing in steel instead of wood.

In approving in 1962 the expenditure of some £400,000 on the Roneo expansion, the Board entered a caveat that it 'saw no obvious advantage in a complete and permanent integration of the interests of the two companies'. For W. D. Opher this was a disappointing attitude for he believed that the agreement could lead to an alliance of great benefit to Vickers at a time when new commercial products were being sought. He persevered, however, and a further link was forged with Roneo when in 1963 Rowe Industries, of Liverpool, sought to take over the manufacture of steel partitions for Roneo. Since Rowe were willing to sell part of their equity, Roneo and Vickers agreed to buy 25 per cent each. The name of the company was changed to Roneo Vickers Partitions Limited, and two years later the remaining 50 per cent was jointly acquired. This was in 1965, and the following year Vickers made an agreed take-over bid for the whole of the Roneo business and so assumed a leading role in the British office equipment industry.

While these developments were taking place in the southern works, the position in Barrow and Newcastle seemed relatively stable, though certainly not without its alarms and excursions.

At one time the existence of Elswick Works had seemed in danger because no major new orders could be foreseen for the Armament Division, its biggest component. A scheme had accordingly been prepared for either closing other activities at Elswick or moving them to Scotswood, in which case the whole of the Elswick site would have been available for disposal. Drastic surgery was avoided, however, by an order in 1963 for 138 Abbott self-propelled guns followed by one for 190 Chieftain tanks. Not least of Vickers 'oddities' since the war has been the recurring conviction that the Elswick Armament Division must be on its last legs: in fact, despite justifiable worries, the Division has constantly been one of the liveliest and most rewarding of the engineering businesses.

Though the overall situation at Elswick had improved, some streamlining went ahead, including modernisation of the foundry as part of a scheme to concentrate foundry facilities in the engineering group at Barrow and Elswick. Efforts were also made to find new products for Elswick and Scotswood. Earlier a licence to manufacture Scott rotary printing presses had brought a useful volume of work to Elswick, and in 1951 a Printing Press Division was formed. An interest in refuse disposal plant was now established with an agreement on the manufacture and marketing of Seerdrum pulverising equipment, based on a design originally developed by William F Rees Limited, of Woking. This venture had considerable teething troubles, however, and failed to realise the hopes pinned on it. More immediately rewarding was a later agreement with Pacific Coast Engineering Company, of California, designed to manufacture and market in Europe the Paceco range of container handling equipment. To this end a new company, Paceco Vickers Limited, was formed, and got off to a good start with orders for Felixstowe and the Port of London. Manufacture was initially at Palmers Hebburn, but later moved first to Elswick and then to Scotswood. This UK link with Paceco followed one developed earlier in Australia with Vickers Hoskins.

While the other engineering works went through their various traumas, Barrow Works maintained a generally high order book and contributed most of the profit earned by the Engineers. In 1961 it topped the £2m mark and repeated the achievement in 1963. Not until the later 1960s did the profit figure fall below £1.5m. These respectable results stemmed mainly from the basic work load provided by the manufacture and installation of ships machinery and boilers and by defence contracts, reinforced by the massive orders from British Rail for Sulzer diesel traction engines. Towards the end of 1964 the Works completed its 1,000th engine for British Rail and a special ceremony was arranged for its handing over to Dr Beeching (later Lord Beeching), then the BR Chairman. Barrow had to be aware, however, that the programme was nearing completion and that a sizeable gap would begin to appear in its order book in 1966. Defence contracts included missile tubes for the Polaris submarines and other work associated with the nuclear submarine programme. Not least, several years of concentrated effort were involved in setting up the Dounreay nuclear facility which was formally handed over in April

1965. A defence activity of continuing importance was the design and development of missile launching systems. Non-defence work at this time included a 17,600-horsepower marine diesel engine for installation in *Australian Star*, the largest such engine built at Barrow. Cement plant and industrial pumps were other important Barrow products, but both were 'fading', due probably to lack of development effort.

While the operating businesses coped with the daily problems of making and selling in difficult trading conditions, Vickers House sought to strengthen further the machinery for group planning and consultation and for co-ordination of recruitment and training at management level.

In much the same way as Lord Knollys and Sir Charles Dunphie had worked together in an *alter ego* relationship, so now Sir Charles, as Chairman, worked with Sir Leslie Rowan as Managing Director. Sir Leslie had been appointed Managing Director at the beginning of 1962, with A. H. Hird becoming Assistant Managing Director.

If Sir Leslie had come to Vickers with an established reputation in the 'outside' world, 'Sandy' Hird was almost the archetypal Vickers 'inside' man. He had been educated at Barrow Grammar School, joined Vickers-Armstrongs at Barrow as an engineering apprentice and took his engineering degree at Imperial College. His professional competence and ebullient personality quickly led to his identification as a 'flier' by Sir Charles Craven, then Vickers chief executive, and when Craven went to the Ministry of Aircraft Production as Controller early in 1940 he took Hird with him.

At MAP Hird made a name for himself as a man who got things done without wielding the big stick and without losing friends. He organised against the clock production of the Barnes Wallis bouncing bomb, with notable contributions by Barrow Engineers, Crayford and English Steel, and then as Director of Machine Tools removed the obstacles that had been delaying the manufacture of bomber undercarriages. When in 1943 Barnes Wallis designed Tallboy, to deal with the V-1 launching sites and E-boat pens, Hird was the man charged with getting it into production as a matter of the utmost urgency. English Steel Corporation again played a part, as did the Vickers-Armstrongs Works at Elswick and Openshaw, and production quickly reached 30 bombs a month. To supplement production in Britain arrangements were made to manufacture Tallboy in the United States. J. G. Lloyd, who was working with Hird in MAP, tells of the dismay caused when the first 25 bombs produced in America arrived at an RAF base in Lincolnshire and it was found that the nose section had been fabricated rather than constructed as a steel machined casting as the specification required. Lloyd was despatched forthwith to investigate and was able to report that, machined or not, the welding was perfect and the bombs need not be scrapped. Following Tallboy, Barnes Wallis designed the 10-ton version known as Grand Slam and again the Hird team took responsibility for its production.

Hird returned to Vickers in January 1945 as Vickers-Armstrongs' Sales Director (bringing with him J. G. Lloyd), and this was still his role when he joined Vickers Limited Board in 1958. He had worked prodigiously in the years after the war to fill empty capacity while new commercial products were developed. In particular, he was largely instrumental in acquiring licences in the United States for the manufacture of petrol pumps, power presses, letterpress printing machines and Scott newspaper presses. It was he, too, who recommended purchase of the Worssam bottling machinery business, and the manufacturing arrangement with George Mann & Company Limited which led eventually to the Mann acquisition.

Obviously performance, personality and seniority played their part in his promotion to Assistant Managing Director, but possibly it represented also a concession to those who alleged that Vickers was excessively preoccupied with production. Unhappily, the appointment was to last less than four years because of a break-down in health, culminating in his death in the autumn of 1965. Some weeks later one of the non-executive directors, Sir Sam Brown, also died. He had been a director for much the same period as 'Sandy' Hird, bringing to the Board legal experience which had been valuable not least in a study of the position of ESC in relation to possible re-nationalisation.

With Sir Leslie's appointment as Managing Director in 1962, the vacated post of Director of Finance went to R. P. H. Yapp who also took responsibility for oversight of major contracts. The 1962 changes did not involve new appointments, and it was not until 1965 that new faces appeared in the Board room. Sir Peter Runge joined in September 1965 as a non-executive, and in November J. H. Robbie became Director of Finance when Mr Yapp took over as Assistant Managing Director following the death of 'Sandy' Hird. The recruitment of Sir Peter Runge was seen as a notable addition to the non-executive strength of the Board. He was a vice-chairman of Tate & Lyle, and additionally played a very positive role in corporate activities in industry and the City, notably as President of the Federation of British Industries from 1963–65: a man much liked by reason of his warm personality and quick sense of humour. Jim Robbie, a chartered accountant and a Scot of Scots, was a long-serving Vickers executive who had joined the Company in 1931 at the age of 24. His earlier appointments had given him experience in all three of the main Vickers-Armstrongs activities – aircraft, engineering and shipbuilding. Like many others in the upper echelons of Vickers management, he was essentially gregarious, though capable of 'blowing his top', and he gave strong support to the Company's social activities, particularly the Group golf tournaments in which he was himself a powerful performer.

The infusion of new blood represented by the Runge and Robbie appointments was reinforced through the device of 'Special Directors' which enabled three other men to 'emerge' at the top in 1965. These were Leonard Redshaw, now the Shipbuilders' Managing Director, A. M. Simmers and A. P. Wickens. Alan Simmers, a chartered accountant, had spent much of his career with Firth-Vickers Stainless Steels and English Steel Corporation, but came to Vickers in 1961 as Secretary and Controller of Administration: on becoming a Special Director of Vickers Limited his close knowledge of ESC was turned to account by appointing him also a Director of the Corporation (now approaching re-nationalisation). A. P. Wickens originally joined Vickers in 1954 to manage the tractor project. As a professional engineer, he had held important appointments both in private industry (notably as Chief Designer, Nuffield Mechanisations Limited), and in Government service, and he was Director, Armament Design Establishment, when he joined Vickers. Following closure of the tractor project, his responsibilities became focused on liaison with the overseas companies, leading to his appointment as Controller, Overseas Department, working to 'Sandy' Hird. In effect, he had now taken over this part of the Hird portfolio. Later, on his appointment to the Board, he became Director in charge of Overseas Interests, a task he performed with relish. 'Wick,' commented a colleague, 'has more friends in more countries than anyone else in the Group.'

Before achieving director status, Robbie, Simmers and Wickens had all been 'Principal Officers of Vickers Limited', a category of appointment instituted in 1961 in reflection of the wish to strengthen co-ordination of activities and practices within the Group. At that time the appointments comprised the Comptroller (J. H. Robbie) and the Controllers of Research (Dr C. F. Bareford), Contracts (N. H. Jackson) and Forward Development (J. G. Lloyd). It was then, also, that a Group Overseas Department had been set up 'in view of the growing importance of the overseas interests in the Group'. Shortly afterwards a Controller of Education and Training was appointed. This was T. S. J. Anderson, CBE, Director of Studies at the Royal Military Academy, Sandhurst, a sociable man of high ability, quietly persistent and persuasive. The appointment gave further impetus to the development of manpower policies on a group basis, and in recognition of wider responsibilities Anderson was later given the resounding title of Controller of Manpower Development, with three Assistant Controllers, for education and training (C. L. Old), employment policies (A. L. Cooper) and staff development and deployment (Glyn Hughes).

Enter McKinsey 1964

When he became Chairman in 1962 Sir Charles Dunphie had no hesitation in deciding to continue the annual group management conferences instituted by Lord Knollys, and conferences were held at Stratford-upon-Avon in 1962 and 1963.

In his opening address at the 1962 conference Sir Charles spoke of the problems involved in achieving a right balance between control at the centre and autonomy in the operating businesses. Clearly he and Sir Leslie felt that the centre needed to take a firmer grip, and in this they were supported by the non-executive directors who were worried by the disparity between forecasts and performance. The setting up of a system of Principal Officers to strengthen the organisation at Vickers House represented a step towards this end, and the Principal Officers were allotted an important role in the 1962 and 1963 group conferences.

'It is essential,' said Sir Charles, 'that a Group such as Vickers should have control at the centre. More autonomy has recently been introduced among the subsidiaries, but the centre must keep a careful watch over general policy, finance and important personnel matters, and it is important not to allow this trend towards autonomy to go too far.'

He also asserted firmly the need for the centre to provide 'certain general services', even if a need for these services was not felt by the bigger subsidiaries.

A principal topic at the 1962 conference was the need for 'simplification of organisation and procedure' in order to free senior managers from 'some of the clutter of daily work' and to give them time to think ahead. There should be 'ruthless elimination of the inessential', said Sir Charles.

As a result of this debate two committees were set up, one to look into Vickers House procedures and the other to do so in the Group as a whole. A year later, at the 1963 conference, the committees reported back. Something had certainly been achieved, but Mr Yapp found it necessary to emphasise that there could be 'no lessening of the need for cash forecasts': these were an essential tool of management, 'especially at the present time'. In fact, he said forthrightly, more forecasts might be asked for 'during the next year or so'.

In saying this he was recognising an obvious dilemma. Firmer control at the centre must inevitably mean more administrative machinery, as already evidenced by the Principal Officer structure, and more paperwork to ensure the flow of information to Vickers House. Everything hinged on how one defined 'inessential'.

This reference to the possibility of more forecasts being required had also to be seen in the context of an announcement by Sir Charles in opening the conference. It had been decided, he said, to engage McKinsey & Co 'to see if the machinery of control is right'. McKinsey & Co, he went on to explain, were 'an experienced American firm which had recently advised other large British groups, including Shell, Dunlop and ICI'.

McKinsey were at that time approaching the peak of their fame as the supreme management 'witch doctors'. Management consultancy had become high fashion and the financial press extolled its virtues, as later in the 1960s they were to extol the virtues of the financial 'whizz kids'. The image of British industrial management was said to be hidebound and fuddy-duddy and the sooner that systematic planning and streamlined procedures could be introduced on the American model the better it would be for all of us, especially now that the computer could be summoned in aid. Feeling no doubt that they must at least see if there was anything in it—possibly, too, as a purely defensive ploy—many companies jumped on the band wagon and the management consultants flourished as never before. Sceptics were not wanting, but at least industrial management was put under compulsion to think hard about its structure and methods, and much more good than harm must have resulted.

So far as Vickers were concerned, structure and methods had been under conscious scrutiny for at least five or six years. It had, indeed, been inescapable given the changes enforced on the Company by the necessity to build up commercial products and by the BAC merger. Nor could it be said that in some major areas at least the Company was failing to keep pace with advanced management techniques. The nuclear submarine programme at Barrow required use of the most sophisticated methods of production control, and computers had been installed at Newcastle as well as at Barrow. The fact remained that whatever may have been done to rationalise, to streamline, to reduce paperwork, to improve communication and to strengthen management,

7

the Group was failing to generate adequate profits. To some extent this might be due to extraneous factors over which the Company had no control, but clearly every effort had to be made to identify and remove internal weakness. To call in management consultants of proved competence therefore seemed a logical and sensible move.

The formal announcement of this decision was made in January 1964. The purpose in seeking an outside opinion, it said, was 'to point out further areas for improvement, particularly in the overall organisation and planning and control of the Group'. The announcement added that the inquiry was expected to last between three and five months. This timetable was achieved and by July 1964 firm recommendations had been made and accepted, and action initiated.

The basic problem, it was recognised, was inability to improve and expand existing businesses and to enter new businesses because of the low return on investment. Since 1959 capital expenditure alone had exceeded net cash flow by over £20m. To meet these expenses, and to finance work in progress, a major increase in borrowing had been necessary, and in 1963 total borrowing stood at £68m, an increase of some £23m since the end of 1962. The increase would not have been necessary had the operating assets been able to earn between 12 and 15 per cent on capital, a target no more than equal to the average return in many UK manufacturing industries. In fact, the Group's return on capital had fallen from 8.8 per cent in 1960 to 5.5 per cent in 1963. With a return as low as this there was little scope for obtaining growth through internal financing.

'The market price of our shares,' it was noted, 'is only 50 per cent of the book value represented by the assets behind each share. This makes it extremely expensive either to raise outside capital or to acquire other companies. Doubling our profits would, in effect, cut in half the cost of any acquisition made through a share exchange.'

If this diagnosis contained no surprises, neither, in general terms, did the statement of objectives. To achieve 15 per cent return on capital would require profits tripled from their 1963 level. This called for 'more forceful and fact-founded management throughout the Group'. It called also for increased efforts to move into 'attractive and growing fields'. 'Our dependence on the armament business is as high or higher than at any time in the last four or five years (32.1 per cent of total Group productive wages). This does not mean that we want to reduce our armament business, but only its significance in relation to the total.' A further objective must be to 'increase our efforts to anticipate both major problems and opportunities and to take positive action to deal with them'.

All this, McKinsey advised, added up to a need for a new approach or philosophy in management and control. It required 'a more formalised, factual and disciplined way' of running the Vickers businesses.

The means to achieve these impeccable ends, McKinsey further advised, was the Corporate Plan — again not a startling new thought since it had already been mulled over in Vickers House and advocated at the 1962 management conference by J. G. Lloyd, as head of the Forward Development Department.

The merits of an annual corporate plan, it was argued, lay not only in stating precise targets for each business: it would force management to anticipate major problems and opportunities, and also to look for major opportunities together with a timetable for achieving them. Moreover, it would allow responsibility to be delegated to individuals. Indeed, it was the essence of the matter that responsibility for success or failure should be pinpointed on individuals, a well-established tenet of McKinsey doctrine.

The Corporate Plan was seen as being based on annual management plans prepared by each business. Among other things, each plan would set out both immediate and longer-term financial goals: it would identify opportunities and problems: it would recommend action designed to expand the business either through acquisition or internal development: and it would state the financial implications of such action in terms of profits, return in investment and capital requirements.

As each management plan was prepared it would be submitted to the executive director responsible, and it would be his duty to evaluate the plan and to challenge it as he thought necessary, with a view to determining priorities between activities and deciding which businesses he would recommend for 'expansion, maintenance or elimination'. In this way the executive directors would consolidate plans for the businesses for which they were responsible, and the consolidated group plans would then be brought together and evaluated in Vickers House.

Once its management plan had been approved, each business would be required to submit a quarterly assessment in order to highlight, in good time, any respects in which the plan was not being met and to explain why this was so.

Finally, at the end of the year, performance would be reviewed against plan and management judged on whether it had attained or improved upon its objectives.

To implement these procedures, some organisational changes were required and these were announced in July 1964. Having declared the intention to introduce 'a new management planning procedure throughout the Vickers Group', the announcement said that a Group Planning Department was to be established at Vickers House. The new department would absorb the staff of the existing Forward Development organisation. Its main responsibility would be 'to advise, in conjunction with the Comptroller, Mr J. H. Robbie, on the formulation of objectives for the Group and its operating companies, and to co-ordinate the individual management plans'. To head the new department, as Controller of Group Planning, Peter Scott Maxwell was brought from Barrow Engineering Works, where he had become Deputy General Manager following his earlier experience on HTP submarines and as managing director of the Instruments business. Like the other Principal Officers he was directly responsible to Sir Leslie Rowan.

To facilitate the management planning procedure in the engineering businesses it was also announced that an Engineering Group was to be formed, comprising not only the activities of Vickers-Armstrongs (Engineers), but also certain other engineering businesses, such as Robert Boby and George Mann, which had been reporting separately. W. D. Opher was to be responsible for the Engineering Group, and he would be assisted by a headquarters staff including Robert Wonfor (Marketing Director), E. P. Tomlinson (Commercial Director), S. P. Woodley (Manufacturing Director) and J. G. Lloyd (Administrative Director). Opher continued to have his own main base at Barrow, but three of the four directors operated from Vickers House. All these changes took effect on the 1st August 1964, and the new planning procedures were set in motion immediately to enable a corporate plan to be prepared in 1965.

In January 1965 a further announcement said that the businesses of all the wholly-owned Vickers-Armstrongs and other UK subsidiaries were to be transferred to Vickers Limited with effect from the beginning of April. This decision was taken by the Board on the grounds that it would simplify the Vickers structure and facilitate the new management planning procedures. In arguing for the change Sir Charles emphasised that Vickers-Armstrongs Limited now performed neither operating nor administrative functions, the administrative role having been taken over by the long-established Management Committee at Vickers House under the chairmanship of the Managing Director. The one possible disadvantage was that the present members of the boards of the subsidiaries might feel that they had lost status through removal of their statutory backing, but in fact they would retain their existing appointments and responsibilities.

The Board accepted these arguments, and Sir Charles wrote a

personal letter to all employees, dated 20th January 1965, in which he explained that the change was an administrative one 'designed to streamline the organisation in order to make it more efficient'. They would now become employees of Vickers Limited and each would receive a formal offer of employment from Vickers Limited 'with the same rate of pay and conditions and with all the rights which you have now'.

In his letter Sir Charles also announced a new sickness benefit scheme, with the whole cost borne by the Company. This scheme was designed to supplement the existing Group pension and life assurance scheme.

In the meantime, preparation of the corporate plan for 1965 had gone ahead, not without difficulty given the novelty of the procedures and the tightness of the timetable. As put before the Board in January it seemed full of good cheer. By the time the half-year results had been put together it was seen to be riddled with miscalculations. 'It is now clear,' the Board was told in September, 'that the management plans of the three main groups were based on incorrect and over-optimistic assumptions.' In fact, profit before tax for 1965 fell some 40 per cent below the planned figure.

One of the principal purposes had been to place responsibility for attaining targets on named individuals. Were heads now to roll and if so whose? Were the managing directors of the individual businesses to take the rap or should it be the chief executives who had approved the miscalculated forecasts? Or those in Vickers House who had set the targets?

In fact, given that the 1965 plan had been a first attempt, and to that extent experimental, the task was less to pinpoint blame on individuals than to assess what had gone wrong, and why, so that the lessons could be applied in preparing subsequent plans.

This was one of the subjects on the agenda of a conference of Vickers Limited directors and principal officers held in October 1965 at 'Bankfield', the Shipbuilding Group's guest house near Barrow. One decision had already been taken—that in future Vickers House would not set targets. The conference noted that, in general, the smaller operating units had come closer to their targets than the large units. This was attributed to four principal reasons—the more intimate knowledge possessed by the chief executives of the smaller units of works conditions and markets: more satisfactory labour relations in smaller units: a proportionately greater technical strength: and closer contact between the various echelons of management.

All these were reasons for creating smaller autonomous units to the extent that this was possible, especially in large works housing several activities and providing various common services to them all. The Bankfield conference firmly approved the principle of smaller units, and indeed the Engineering Group had already put in hand the formation of several such units. Later it was to be felt that the enthusiasm for smaller units had been excessive, and some were returned to a works setting.

The conference was also asked by the Chairman to consider whether there was a need to give greater rewards for success and less security for failure. Vickers, it was thought, had become widely regarded as an employer providing absolute security of employment, irrespective of performance. Here again the conference took a clear view—that achievement must be rewarded, but that more generous treatment for successful employees must be accompanied by stricter measures towards those who did not succeed. If dismissals became necessary, however, fair severance payments should be made.

These were issues which would no doubt have been raised irrespective of the shortcomings of the 1965 corporate plan, but the plan had served to highlight them. It had also helped to intensify thinking on 'whither Vickers', both in general terms and in relation to particular activities. Certainly there was no disposition to suggest that a system of annual management plans should not continue. Its value as a framework of discipline and a focus for regular re-appraisal was not in doubt. The immediate task was to improve the quality of the 'feed-in', and this required consideration of psychological factors as well as competence in the techniques of management. It also had to be recognised that, however well a plan was produced, there would always remain a number of imponderables which at the end of the day might completely invalidate its conclusions, particularly those reaching beyond the year or two years immediately ahead. The importance of a planning system lay more in its disciplined procedures than in its ability to produce infallible forecasts and conclusions.

3

Pastures New 1966–70

For the last two years of his chairmanship Sir Charles might reasonably have expected some relief from the hard going of the first three. In fact, his worries were to multiply. In one respect, however, 1965 and 1966 brought satisfaction. With the VC10 no longer to be financed, and with a sharp reduction in borrowings forecast, Vickers were at last in a position to put resources into building up those businesses in which growth could be anticipated. Moreover, the prospect now existed of a considerable infusion of cash from the re-nationalisation of ESC. This did not mean that re-nationalisation would be welcomed, but the new Labour Government had firmly declared its intentions and short of a most unlikely political upheaval the loss of ESC had to be expected. With the payment of compensation, Vickers would not only have a large amount of cash for disposal, but would also be under strong pressure from its stockholders to say quickly how it would use the money.

The conviction already existed in Vickers House that the Company should seek to establish strong positions in the printing machinery industry and in medical engineering. The printed word, it was argued, would always be with us, and the demand for printing machinery must grow, especially in the developing countries: true, American and German competition was intense, but Vickers was already in the business and had the experience and know-how which would enable it at the very least to hold its own in a growing market. As to medical engineering, the development of health services throughout the world must create a rapidly expanding market for hospital equipment: true again, competition was intense, especially American competition with the advantage of a vast domestic market, but the work done over the years by Vickers Research had put the Company into the forefront of oxygen therapy equipment, with automated laboratory equipment of a highly-sophisticated kind also far advanced towards production.

Here, then, were two areas ripe for expansion. To some extent the means were available, but the opportunities had also to be found or to be created. In fact, they presented themselves.

In January 1965 it became known that Oxygenaire Limited would accept a take-over bid if satisfactory terms could be negotiated: though not a large company, Oxygenaire had a considerable reputation for its marketing expertise in medical equipment, and this was precisely the expertise Vickers medical engineering now required.

A few months later, largely through good personal relationships established with Mr Charles Crabtree, chairman of R. W. Crabtree & Sons Limited, it was indicated that he and his colleagues felt that the time had come for his company to move into a setting in which more resources would be available to support and develop the printing machinery business which the Crabtree family had so successfully established at Leeds: nothing could have suited Vickers better for here was the opportunity to acquire a business with a high reputation throughout the world, a business operating moreover almost cheek by jowl with Vickers own printing machinery subsidiary, George Mann & Company Limited.

Both negotiations went ahead satisfactorily, aided by an injection of cash resulting from Vickers Australia 'going public' to the extent of one-third. Oxygenaire was acquired for £601,000, with net assets assessed at £257,000 and goodwill £244,000. The Crabtree acquisition cost £7m (net assets £5.1m and goodwill £1.9m).

Nor were these the only acquisitions made in 1965. Vickers finally and completely committed itself to the chemical engineering industry by acquiring for £500,000 the remainder of the issued share capital of the Zimmer companies, following the untimely death of Hans Zimmer.

By the middle of 1966 Vickers had also committed itself to the office equipment industry by acquisition of Roneo Limited, despite the Board's earlier doubts about the wisdom of going much beyond existing arrangements with Roneo in the manufacture of steel furniture and partitions. Roneo, like Crabtree, had been prompted by the belief that it could not itself generate the strength required to maintain growth in conditions of intense competition. The acquisition took effect on 1st July 1966, for a price of £7.12m, of which £2.53m represented goodwill. Just over half the cost was met by issue of Vickers shares with a premium of 6s 9d per share.

To round off the shopping list, further additions were made in 1966 to printing machinery activities by the acquisitions of the Otley firm of Waite & Saville Limited, a small but well-established business producing sheet-fed offset presses, and of Dawson, Payne & Lockett, a service organisation established in London.

Within little more than a year Vickers had therefore spent nearly

£16m on acquisitions, and declared its choices of printing machinery, office equipment, medical engineering and chemical engineering as the main new growth areas.

Not in all instances had the Board approved these choices without reservation, and in the instances of Zimmer and Roneo it might have been felt that events had forced the issue. The case for medical engineering had been argued with some force from the executive side, but one non-executive director at least had expressed misgiving by emphasising that it must be given a clear target and not allowed to fail. Perhaps the expansion in printing machinery went forward with the greatest degree of general approval. Time had now to show whether basic assumptions had been right, and whether the new acquisitions could be successfully absorbed and integrated.

Employees in the Crabtree companies numbered over 3,000, and with nearly 600 also working for George Mann & Company, printing machinery now ranked as a major Vickers activity. Combined sales in 1964 had approached £10m and it could be claimed that Vickers was 'now the leading British manufacturer of printing machinery'.

The Crabtree group comprised a somewhat diffuse empire, both in product and location. Its products ranged from massive newspaper printing presses, requiring up to two years to design, build and install, to lithographic printing plates for instant delivery. It had factories not only at Leeds, but also at Gateshead, Luton and London, some very ancient, some very modern. The business began in Leeds in 1895, not in printing machinery but in textile machinery and general engineering. Printing machinery first appeared among its output in 1905, but with the success of its first newpaper press in 1920 it pinned its fortunes firmly to products for the printing industry. Founders of the business on a capital of £200 were Mr Richard Wadsworth Crabtree and Mr Alfred Horsfield. The young Charles Crabtree, Richard's son, began work there on the opening day and presided over its later expansion, largely by the acquisition of other family businesses.

The Mann company, acquired by Vickers in 1947, had an even longer history and was due to celebrate its centenary in 1971. Its founders were George Mann, a fitter, and Charles Pollard, a pattern-maker, who came together to build a lithographic printing machine of their own design. The story is told that when the time came to give the infant company a name George Mann suggested that they should toss a coin to decide, but Pollard would have none of it. 'Nay, George,' he said, 'put thine down. They'll come first to thee for the rates.'

Included in the Crabtree group was W. H. Howson Limited, which had been acquired cheaply in 1951 as a go-ahead manufacturer of deep-etch chemicals for lithographic printing plates. Here again it was a story of small beginnings. The prime mover was Hector Howson, a letterpress printer, who during the war, worked as manager of the print shop of a Leeds pharmaceutical manufacturer and had become interested in the chemical processes of making deep-etch printing plates. He decided to try his hand at producing deep-etch chemicals himself and did so, first in the cellar of his home in a terraced house in Leeds, and then in a stable owned by his twin brother, a potato merchant. Since his own expertise was in letterpress printing he took into partnership Charles Hunter, a skilled platemaker, and also drew into the business Gilbert Matthewman, a colleague from the pharmaceutical firm who could contribute chemistry know-how. The company was set up in 1946 and at once began to thrive, to the extent of attracting export as well as domestic customers. Its operations were still on a small scale, however, and even in 1950 it had only eleven employees, though this did not inhibit grandiloquent references to its No 1 and No 2 factories—No 1 comprising three terraced houses and No 2 a Nissen hut. The first contact with Crabtrees arose when Howson decided to set up a London office and heard that Crabtrees had space available there. The immediate Crabtree response was an offer to buy the business. Howson was influenced by the fact that he had a heart condition and was afraid that if he died suddenly his wife would be left with a business she did not understand and could not run. He had, in fact, already offered to sell to Algraphy Limited, but had been turned down. The Crabtree offer therefore looked attractive to him and he accepted without more ado. Had Algraphy taken the opportunity presented to them, the history of lithographic plate manufacture in Britain must have taken a very different course for in 1969 Algraphy was itself acquired by Vickers to be merged with Howson.

Under the Crabtree aegis, the Howson business was given every encouragement to add plate-making to its production of deep-etch chemicals, and in 1954 it produced the first commercially successful anodised plate. Technological success, founded on a strong research effort, was reinforced by extremely vigorous selling policies, spearheaded by F. P. ('Bob') Tanner, a man of remarkable drive and determination, who joined the company in 1957 and became sales manager in 1959. In 1957 the business moved to a brand-new factory at Seacroft, an industrial development area on the outskirts of Leeds, and in this setting it devised highly efficient production methods under the direction of Derek Varley, who had joined Howsons in 1950 as a laboratory assistant and was now works manager. In 1962 a subsidiary company was set up in Soest, Holland, to manufacture Howson products for the EEC countries. By 1964, the year before Vickers acquired the Crabtree group, Howsons had a turnover of nearly £1.5m and in profit-making had taken a lead over the printing machinery divisions.

In buying Crabtrees to establish a leading position in the UK printing machinery industry, Vickers had thus acquired as a bonus a very lively stake in the printing plate industry. Fortunately, the Howson potential was quickly recognised, and the thrusting team at Seacroft found very ready backing for their ambitious plans. 'Bob' Tanner became managing director in 1967, and in 1969 Vickers acquired Algraphy, the principal UK competitor, to form Howson-Algraphy Limited. In due course, Howson-Algraphy was to be hived away from printing machinery activities to become an operating group of Vickers Limited in its own right. As such it was to outstrip all other groups as a profit-earner. Ironically, as printing plates prospered, printing machinery struggled and lost its own status as an operating group.

Success in industry, it would seem, derives as much from chance and the ability to seize opportunities as from plotting careful courses.

The Algraphy acquisition in 1969 cost Vickers £7.9m. At the time there were those who thought the price unduly high, but in the perspective of time and results it was amply justified. Algraphy had been in business since 1898 when it was formed in London to exploit the patents on the first aluminium lithographic plates in Britain. As sales increased, the manufacturing location was moved several times, though always in the London area, and in 1969 the main factory was at St Paul's Cray, near Orpington, with a second factory at Margate for the production of small offset plates and chemicals. As with Howson, Algraphy had set up a manufacturing subsidiary in Europe to supply the EEC countries, their choice of location being Milan, so that the merged company had factories in Holland and Italy. Both companies put considerable effort into exports, and Howsons could claim that 40 per cent of their production went overseas to nearly 100 countries. Together they employed over 1,000 people and stood among the world leaders in the printing plates industry.

As a result of the Crabtree acquisition in 1965 a new figure emerged in the upper reaches of the Vickers hierarchy as chairman of the newly-created Printing Machinery Group. This was Colonel H. S. J. Jelf, CBE. His military title stemmed from some 20 years in Army engineering, including a senior command in Pakistan, but he had impeccable Vickers qualifications, as an Old Barrovian and a Barrow apprentice. He rejoined Vickers in 1957 and two years later became managing director of the George Mann business. He was thus well versed in the printing machinery business, and his appointment as chief executive of the combined companies could have caused no surprise. Headquarters were naturally enough in Leeds, the mother town of both companies, and the Crabtree name remained in the title, Crabtree-Vickers, with Charles Crabtree as president (he died towards the end of 1966 at the age of 85). The third generation of the Crabtree family was represented by Peter Crabtree, who was to remain an executive director until 1969. Also coming to the fore, as a result of the acquisition, was R. O. ('Ron') Taylor, a Scot who had been appointed chief accountant of George Mann at the age of only 27, and then commercial manager and secretary: he now became a member of the Crabtree board, as well as the Mann board, and later, still in his thirties, he was to be managing director of the Printing Machinery Group as a whole. With the formation of Howson-Algraphy he took on the role of deputy chairman of that company, in addition to being managing director of Crabtree-Vickers, and when in 1973 Howson-Algraphy was set up as an operating group in its own right he became its chief executive, with F. P. Tanner as managing director. In 1977 he was further appointed a Director of Vickers Limited.

Following the acquisition of the Crabtree group came the successful bid in 1966 for Roneo Limited, and in July Vickers was able to announce that a Roneo Vickers Office Equipment Group was being formed under the control of a reconstituted Roneo Limited Board. As a result of the acquisition, said the announcement, Vickers had 'established a position of major strength in this important and growing activity, and from this position it will seek to expand vigorously'.

As recently as the conference at Bankfield in October 1965 the view had been taken, despite the arguments of W. D. Opher, that no interest should be acquired in Roneo on the grounds that it would mean involvement in the manufacture of business machines and equipment whereas Vickers interest lay in metal furniture and partitions. The existing association, it was felt, met all Vickers purposes and was secured by a 10-year agreement. The *volte face* a few months later came mainly because of nervousness about the possibility that Roneo intended to sell out, and that the agreement with them on the manufacture of furniture would thus be put in danger. The Roneo board did, in fact, believe that in the face of intensified competition from the international giants they needed an injection of resources which they themselves could not generate. They were therefore exploring the possibilities of tying up with a major industrial company and Vickers was an obvious choice of partner, given the arrangements that had existed for a considerable number of years.

The Roneo business dated from the beginning of the century. Its principal founder was Augustus David Klaber, born in London in 1861 after his parents had emigrated from Prague. In his early twenties he had a job as a stationer's assistant with a firm in Chancery Lane, but he quickly decided that New York was likely to offer more opportunities to a young man of ambition and ability and he opened a small business there in office supplies. Among other things he became an agent for David Gestetner whom he had known in London.

With intense American interest in the development of copying machines, Klaber saw that opportunities would arise both to take out and acquire patents. His big opportunity came when Harry W. Lowe, of Omaha, obtained a patent for a 'Duplicating or Stencil Printing Machine'. This was the machine which established the rotary single cylinder duplicator as the copying machine of the future. Klaber succeeded in negotiating with Lowe, first a licence to manufacture and sell a rotary machine based on the patent, and then a share in manufacturing and selling rights in Britain, France and Germany.

Thus armed he returned to London in 1900 to form the Neostyle Manufacturing Company, with a small factory in Petticoat Lane. The word 'Roneo' (derived from Rotary Neostyle) was among a number he registered in 1901 as trade marks, and this became the trade name for the company's products. The chairmanship Klaber offered to William Thomas Smedley, an accountant with knowledge of the City, and Smedley retained this appointment until he retired in 1933. Also involved in the business were Klaber's son-in-law, Augustus Samuel Newmark, and his son Emile. The latter was for a time in charge of a subsidiary set up in France and given the name in 1907 of Compagnie du Roneo. This subsidiary later became independent, but in the fullness of time was to return to the Roneo fold when in 1969, after the Vickers take-over, it was bought for £1.65m.

Even before the First World War the Roneo duplicator had come into such widespread use that in 1914 the Oxford dictionary gave 'Roneo' the status of an entry. It is perhaps, therefore, not as surprising as it might seem that kept in a glass case in the Museum of the October Revolution in Leningrad is Roneo machine Serial Number 44939. The museum was formerly a villa in which the Bolsheviks set up their headquarters during the provisional government of Kerensky, and machine Number 44939 was said to have been used almost continuously by Lenin after he returned to the villa in 1917.

Klaber died in 1915 after an operation when he was only 54, but the business was now large and important, with a major factory at Romford. It went through a rough patch after the war and in the early 1930s was on the verge of collapse. To the rescue came Arthur Chamberlain, whose place of eminence in British industry was evident in his appointments as Chairman of Tube Investments and the Churchill Machine Tool Company, and as a director of GEC, though his interest in Roneo arose through his membership of the Midland Bank Board. Chamberlain agreed to become chairman and told the staff he would need five years to put things right.

To help him do so he brought in his son Paul. With the outbreak of war Paul Chamberlain went into the Forces, but he returned in 1946 and became managing director in 1947. The chairman was now Sir Greville Maginnis, and the Maginnis–Chamberlain era continued until Sir Greville's death in 1961. At the age of 47 Paul Chamberlain then became chairman as well as managing director. Though 'Roneo' had become a household name, and though the business was producing moderately good results, Paul Chamberlain and his fellow directors were increasingly aware that Roneo lacked the resources it needed for expansion, and that only by expansion could they sustain a competitive position. They felt therefore that a merger or some other form of association must be sought with another firm. In view of the relationship established through Dartford and Roneo Vickers Partitions, Vickers were made aware of the situation and the Vickers decision to bid came in May 1966, spurred on no doubt by the knowledge that a large American conglomerate had begun to show interest.

For Vickers the acquisition represented a venture into the largely unknown. Making steel desks, filing cabinets and chairs they understood well enough, but the Roneo business lived on intensive selling methods, with hundreds of travelling 'reps' ready to take small orders for duplicators and stencils for instant delivery from a chain of depots. 'Brighten us up,' it was said in Vickers, but failure to understand the process caused early trouble. An injunction to Vickers units to do everything possible to cut down stocks was extended to Roneo, and suddenly the 'reps' found that they were taking orders without being able to accomplish quick deliveries, a situation which did a good deal of damage to customer confidence before it could be retrieved.

For a short time after the formation of the Roneo Vickers Office Equipment Group Paul Chamberlain stayed as non-executive chairman, but he retired from the scene at the end of 1966. The new chairman and chief executive was Robert Wonfor, a senior Vickers man whose career included considerable experience in marketing. From St Bees School he had joined Vickers as an apprentice at the Erith and Crayford Works, but then left to take a job in the oilfields of Sarawak and Brunei. After three years overseas he rejoined Vickers, progressing steadily upwards until just after the war he became managing director of the subsidiary Powers-Samas Accounting Machines Limited. From this appointment he moved in 1955 to the sales side of Vickers-Armstrongs (Engineers), becoming sales director in 1956. With the setting up of Vickers Engineering Group in 1964 he was appointed marketing director, and it was from this post that he moved to Roneo Vickers. Like W. D. Opher he already knew the Roneo organisation and its senior managers, and had played a principal role in the acquisition negotiations. He was to remain chairman until his retirement in 1973, and by then he had been a member of the main Vickers board for five years.

During his chairmanship Vickers lived up to the undertaking that they would 'seek to expand vigorously' from the base established by the Roneo acquisition. In 1968 Roneo acquired the share capital of four companies in Australia which had been its agents in Melbourne, Sydney, Adelaide and Brisbane, and also of Hadewe BV, a Dutch company manufacturing folding machines. These were small if significant acquisitions, involving considerations totalling £650,000. In the following year, 1969, two major steps were taken with the acquisitions of Compagnie du Roneo in France for £1.65m and of the Yorkshire firm of Hirst Buckley Limited, a leading British manufacturer of business forms, for £2.9m. Thus in four years, including the acquisition of Roneo itself, over £12m had been spent on establishing a position of strength in the office equipment industry. The strength was based both on range of product and on an international spread of manufacturing and sales subsidiaries. The title 'Roneo' perhaps carried both advantage and disadvantage—advantage because it was almost universally recognised (to the extent indeed of inclusion in dictionaries) but disadvantage because it was immediately associated with stencil duplicators or, at best, business machines in general. Duplicators and copiers still represented a major activity, but almost every other office requirement could now be met with a Roneo Vickers product —metal and wooden furniture, partitions, mail room equipment, business forms. In this diversity of product and location lay a degree of insurance against anything short of international recession, but equally results in any one year could be pulled down by a setback in one product or one location.

Results in these early years were, in fact, encouraging though profits did not increase in proportion to turnover. In 1967, the first full year of the group, turnover totalled £14.1m with a trading profit of £1m. Four years later, in 1970, the figures were £35.5m and £2m respectively. At the 1970 level of turnover office equipment had exceeded shipbuilding and was within a hair's breadth of overtaking engineering.

Figures from Printing Machinery and Supplies were hardly less encouraging—turnover £9.7m and trading profit £0.9m in 1967 compared with turnover £17m and trading profit £2.1m in 1970.

Had it not been for the successful performance of these two new-comers Vickers results for 1970 would have looked very sick since the Engineering Group, Medical Engineering and Chemical Engineering all showed losses. Out of a total group trading profit of £5.1m all but one million came from office equipment and printing machinery.

As was clear from the 1970 results, troubles had accumulated for the two other 'new' areas of activity, chemical engineering and medical engineering.

Following the death of Hans Zimmer and the acquisition of the remaining shares in Hans J. Zimmer AG, the two chemical engineering companies were re-named. The Frankfurt company became Vickers-Zimmer AG and the British offshoot, hitherto bearing the ponderous title of High Polymer and Petro-Chemical Engineering Limited, became Vickers-Zimmer Ltd. The principal purpose of the UK company was to undertake sales in countries behind the Iron Curtain where, in certain instances, notably East Germany, political objections existed to taking contracts with a company based in West Germany. Though Zimmer had customers in many parts of the world, the building of plants in eastern Europe represented an important part of its activities, and it had won major contracts in the Soviet Union and East Germany. Against a background of considerable technological achievement, Zimmer had perhaps become over-confident in tackling the design, construction and start-up of very large integrated plants. Certainly it offered hostages to fortune in one of its Soviet contracts, taken before the full Vickers acquisition. Two plants were being built, one for the manufacture of nylon cord from a substance called AH Salt, and the second for production of this substance. While the nylon plant went ahead satisfactorily considerable problems were experienced in producing AH Salt to the agreed specification and within the agreed timescale. Penalties were incurred, including the cost of providing AH Salt from other sources, and in 1965 provisions totalling £3.1m had to be made. Of this amount, half a million was provided in the profit and loss account of High Polymer and £2.6m by Vickers Limited in the Contingencies Account, the loss being regarded as exceptional and 'unrelated to the year's trading': net of tax the charge was £1.56m. This was not, in fact, the end of the matter for attempts to rectify the plant did not succeed and the contract was cancelled in 1967, requiring provision of a further £2.2m gross (£1.24m after tax) which was also debited to Reserves.

There had thus been an unhappy start to Vickers' involvement in chemical engineering. In the first three full years after the 1964 acquisition it had been necessary to make provisions totalling £2.8m after tax, and trading profit to set against this loss during the three years amounted to little more than £1m. Had the AH Salt failure been an aberration? Zimmer had built scores of plants which had fully met technical requirements. Having got AH Salt out of its system could Zimmer now be expected to go ahead without further serious 'accidents'? 1968 and 1969 produced profit, but at a low level – £0.1m and £0.3m. Then came more trouble. Four major contracts had been taken in East Germany. Two fell seriously behind schedule, in part because of the failure of sub-contractors to maintain delivery dates. Whatever the cause, penalties were now in prospect and in 1970 a provision of £1.55m had to be made which resulted in a loss of £0.9m on the year. Thus in the six years 1965-70 there had been a net loss of £2.3m on Zimmer activities.

Nor was this all. A worrying situation had arisen as a result of the acquisition by Zimmer AG in 1968 of a 90 per cent interest in a polyester fibre business in Southern Italy, the Compagnia Generale Resine Sud spa. The plant had been built by Zimmer AG in 1965 and modernised in 1968 to the point at which it was claimed to be one of the most up-to-date polyester plants in Europe. The case stated for acquiring control was that CGR would provide a valuable demonstration and training establishment for customers, and also enable Zimmer to obtain direct experience in production techniques. An additional factor was the substantial sum still owing to Zimmer for their work on the plant. So far as Vickers Limited was concerned, however, the acquisition was in the nature of a *fait accompli* and raised worrying questions about the relationship between Frankfurt and London. The fact remained that Zimmer had now established a stake in polyester production and had done so, moreover, in a business which was making losses, whatever view might be taken about its future. In short, Zimmer had made a considerable addition to its overheads.

As if all this was not enough, trouble of an entirely different kind had erupted to make headlines in the national press. In September 1967 the Chinese authorities detained two employees of Vickers-Zimmer Limited working on the erection of a polypropylene plant at Lanchow. One of the men, George Watt, was a British subject and the other, Peter Deckardt, a West German. They were held until March 1968 when a Chinese court sentenced Watt to three years'

8

imprisonment on charges of espionage (though, in fact, he served a shorter time than this) and Deckardt to deportation. The incident had nuances which were difficult to unravel, but it caused considerable upset in Vickers House though, in financial terms, not a great deal was at stake. The danger lay in the damage that might be done to future trading opportunities in China, and a strong directive was issued to reinforce the specific requirement in contracts of employment for projects in Communist countries that there should be no embroilment in political activity.

All in all, Vickers was getting little joy from chemical engineering. It was perhaps a classic example of the mistake of moving into a technology in which no expertise existed within the parent organisation. In this instance additional problems in understanding and communication arose from the difference in nationality and language. Vickers attempted to deal with the situation through Vickers representatives on the Supervisory Board, and by appointing a Vickers man (first C. R. Wesson and then L. P. Harrold, commercial manager at the North-East Works) to one of the key management posts in Frankfurt. Before the full acquisition J. H. Robbie was Chairman of the Supervisory Board from 1961 to 1963 when Dr K. Dohrn, a leading German banker, agreed to take over the responsibility, though Robbie remained a member of the Board. With the re-naming of the companies in 1966 as Vickers-Zimmer AG and Vickers-Zimmer Limited, Robbie also became chairman of the London company and Robert Wonfor, deputy chairman. The London managers were, however, made responsible to their opposite numbers in Frankfurt. Robbie retained his Zimmer appointment until 1968 when a series of changes saw J. R. Hendin becoming a member of the Supervisory Board of Vickers-Zimmer AG, and the AG assuming full responsibility for the activities of the London company, of which L. P. Harrold became chairman.

Though Vickers-Zimmer sales showed an upward trend, and reached nearly £23m in 1970, the inability to make adequate profits was obviously worrying. Even more worrying were the extensive guarantees Vickers Limited were being required to make in support of large Zimmer contracts, especially in the light of experience with the AH Salt and East German projects which had shown a disposition in the Frankfurt management to reach beyond technical capabilities. Chemical engineering was now looking more and more like a mistaken venture.

The attempt to set up a major activity in medical engineering was also floundering. Its origins lay in a section of work undertaken by the Group Research Establishment under the direction of Dr Kenneth Williams, the head of medical studies, who had earlier been concerned in the RAF with the use of oxygen at high altitudes. From this experience he had moved into experiments in the therapeutical use of oxygen under high pressure. The range of possible use was thought to be very wide, from baby incubators to high pressure oxygen chambers for the treatment of conditions such as coronary heart disease, carbon monoxide poisoning and post-operative shock. As early as 1958, the year after the Research Establishment had been set up, it was reported that a number of new products had been 'brought to an advanced stage of development', including 'oxygen breathing equipment and flexible headpieces for airmen's pressure suits'. As the work continued, equipment of increasing sophistication was developed and in 1962 the point had been reached when a range of hyperbaric oxygen products could be marketed. 'Vickers,' it was claimed, 'are believed to be the first organisation to offer equipment of this kind for sale as standard products.'

While research and development proceeded at Sunninghill, manufacturing was being undertaken at South Marston, and the time was approaching when a decision would need to be taken about moving these products into a fully commercial environment. It arrived in 1965 when the formation of a Medical Engineering Division was announced, with headquarters in London and manufacturing not only at South Marston but also at Basingstoke as a result of the acquisition of Oxygenaire. The importance of the acquisition was seen in the marketing expertise it brought into the division, for Oxygenaire had built up an excellent distribution system and after-sales service for its own medical products, including baby incubators.

Dr Williams was appointed to take charge of the new division. His plans for its expansion stretched far beyond hyperbaric equipment. At the Research Establishment, he had been involved in other medical research, including tissue storage and automated biochemical analysis. At one time, with transplant operations coming into vogue, considerable hopes had been placed in the development of tissue storage equipment, but for technical and other reasons these efforts did not lead to a marketable product.

Interest became focused instead on the development of an automated biochemical analyser, a machine which in layman's terms would carry out simultaneously and at a very fast speed up to 20 tests of blood plasma and serum.

The successful production of such a machine would obviously have considerable significance for the efficiency, speed and organisation of laboratory work, and its further development became a responsibility of the new Medical Division. The support of the National Research Development Council was enlisted, and in 1967 the Medical Division was able to announce that the Ministry of Health had agreed to buy the first batch of the machines for use in hospital laboratories. In referring to this success in his 1967 report, the Chairman said: 'This machine, the Multichannel 300, is well in advance of any other machine of its kind in the world and its introduction opens up new possibilities for the automation of laboratories.' He added: 'Its purchase by the Ministry of Health demonstrates how Government purchasing power can and should be used in support of British technical innovation and exports.'

On the face of it, here was a major success. Unfortunately, the machine still had 'bugs' and its launching as a product was almost certainly premature. Further heavy development expenditure was still required and in the meantime the 'bugs' had raised doubts about the quality of its performance. Because of pressures to reach the market ahead of competition, a chance had been taken and had not come off. There was more to it than this, however. For a machine as sophisticated and expensive as this the UK could not provide a market large enough to enable development costs to be recovered. Even with the machine installed in all major British hospitals, costs could be recovered only by substantial sales overseas, with the United States as the crucial market. Selling arrangements were set up in the United States and a thorough study made of the market, but resistance to buying non-American equipment proved difficult to overcome, especially in the initial stages when 'bugs' still existed. The delay before the Multichannel 300 became fully efficient gave the principal American manufacturer time in which to develop a competitive machine. Though it could be argued that in certain respects the American system had less satisfactory features, the Multichannel 300 was at a cost disadvantage and could not now make sufficient inroads into the North American market to bring it commercial success. Sales were made there, both of the Multichannel and its derivatives, and also in other export markets, but even with these sales and with the British hospital system fully equipped, the possibility of recovering development costs no longer existed.

The situation was recognised in 1970 to the extent of a decision to write off against profits £750,000 in development costs on the Multichannel 300 and another apparatus, the Vickers Auto-Tape System, designed for cytology screening and other applications, including histology.

In the same year a provision of £400,000 was made to cover an expected loss on the design and building of a hospital at Riyadh in Saudi Arabia. This was in addition to a provision of £375,000 made in 1969. At the time they were taken in 1968 and 1969 the contracts had seemed to provide excellent justification for the action taken in 1966 to establish, as part of the Medical Engineering Division, an 'organisation for equipping complete hospital projects'. As a concept it could hardly be faulted. The less developed countries were anxious to build up their health services, particularly by the building of hospitals and for them it was an obvious advantage if a contract could be placed for the whole project with a single contractor who would take responsibility for design, building, equipping and staff provision and training. For Vickers Medical Engineering such contracting would have the additional advantage of providing sales outlets for the equipment they had developed and were developing. In pursuing these possibilities Dr Williams and his colleagues travelled extensively overseas, but though discussions in some instances reached an advanced stage contracts proved extremely difficult to obtain. The Riyadh project began with a design contract in April 1968 and the main contract for construction followed in December 1969, with Vickers Medical Engineering as consultants and project managers. Design changes as the project proceeded caused a considerable escalation in costs, however, and with these increases not provided for in the contracts provisions against loss became necessary.

The conjunction of these problems and those being experienced with the Multichannel 300 meant that in 1970 Medical Engineering showed a loss of no less than £1.2m. In none of the five years since the division was formed had a profit been made, and re-thinking was clearly necessary. Organisational changes in 1969 and 1970 led to medical engineering becoming a division of the Engineering Group. This loss of independent status was accompanied by disbandment of the hospital projects section, though with continuing arrangements for the Riyadh project. The main ongoing activities were then hyperbaric equipment, hospital gas pipe-line systems and automated equipment such as the Multichannel 300. Dr Williams left the division to become director of the Riyadh hospital with effect from 1st January 1970.

As the early euphoria evaporated, Vickers was left to contemplate the nice balance between ambitious pioneering and commercial prudence. Though medical engineering might have its place in the scheme of things it was certainly not one of the major 'new' growth activities for which Vickers was looking. In 1970 big question marks also hung over chemical engineering. Results from the Printing Machinery Group seemed encouraging, but these owed much to the performance of Howson-Algraphy who were concerned with plates and not presses. There remained office equipment and here hopes remained high.

4

In the Hands of Whitehall?

'A giant in the hands of Whitehall,' wrote a financial journalist. He was uncomfortably near the truth. All industry in the 1960s was feeling the growing impact of Government control and intervention, but Vickers was more vulnerable than most. Not only did it live on the commanding peaks which Labour Governments were determined to capture, but also as a defence contractor it was strongly affected by the twists and turns in defence policy, which took place from government to government and even from Minister to Minister.

Within Vickers the reaction tended to be relatively mild. In the nature of their activities they had long been accustomed to treading the corridors of Whitehall. Power had to be dealt with where it lay. If change was inevitable the sensible thing was to try to make the most of it, or at least to adapt to it. It was no use getting hot and bothered. No use making vehement public denunciations. The task was rather to talk, to persuade and to wield influence in the right places. Of course, Government and industry had to work together. Government had an essential role to play, not least in protecting the interests of British industry in the rough and tumble of international trading. So you maintained a dialogue and tried to make it constructive.

Sir Charles Dunphie touched on the topic in his valedictory speech at the 1967 annual general meeting.

'Under successive Governments,' he said, 'collaboration between Whitehall and industry has steadily grown and this is a development which, in the circumstances of the 1960s, is both inevitable and right . . . Yet in welcoming collaboration I think we ought to take care to see that there is a proper husbanding of the time and effort involved. In effect, this is an additional overhead. The time of senior executives is costly. Let us make sure, therefore, that we use it to best advantage. There is another point. In equipping itself to take on new responsibilities in relation to industry, Whitehall draws on a reservoir of senior executives and technical manpower which is far from inexhaustible. To some extent it must do so at the expense of industry . . . Here again is something which needs to be kept in balance . . . We could quickly find ourselves through the looking glass into a world in which all plan and none produce.'

Later Sir Leslie Rowan went more deeply into the subject when he addressed a seminar on 'Problems in Industrial Administration' at the London Graduate School of Business Studies.

'If one had to identify the factor which more than any other changed the British industrial scene in the last few years, it would be the mushrooming of Government intervention in industrial affairs,' he said.

'Variations in taxation policy are something with which companies have had to learn to live, but perhaps in the Sixties we had more variations and permutations than ever before. Sticks and carrots of many novel shapes and sizes came our way and made company accounting and forecasting more than ever difficult.

'But it is not of this I am chiefly thinking. It is rather the degree to which Government began to intervene in the very structure of industry, setting up agencies, often with nebulous authority and powers, to achieve ends determined by the men in Whitehall, often anonymous. Now, as a former man of Whitehall myself, I am quite satisfied that you will find a good deal of wisdom there, but it is not the whole wisdom; someone said unkindly that it was wisdom "unencumbered by the distortions of practical experience"; more kindly, it could be called wisdom which has not been tempered by direct experience of the harsh facts of the market place.

'Some Government intervention and oversight there must be in the overall national interest, but just look at the proliferation of agencies we have had in recent years – NEDC and all the little Neddies, IRC, SIB, PIB, the Monopolies Commission, NRDC and CIR, not to mention DEP, the Industrial Training Boards and a variety of other direct and indirect interventions by MinTech and the Board of Trade, now combined into one Department of Trade and Industry.

'Think also of the powers exercised by Government as an employer and as a purchaser, often a monopoly purchaser, just as the nationalised industries are often monopoly suppliers. Think of all the statistics that are demanded to feed into the maw of this all-enveloping machine and the amount of non-productive time that has to be used in industry to satisfy its insatiable appetite.'

In judging this issue, Sir Leslie concluded, he was quite clear on five points.

'(a) Industry must not accept that something is in the public interest merely because a Minister says so, unless he is announcing policy approved by Parliament.

'(b) We need to regard with great suspicion both requests for "voluntary co-operation" and also the transfer of powers to parastatal bodies with roles not clearly defined and limited by Parliament.

'(c) We must be careful not to be caught in the inconsistency of regarding Government intervention as acceptable when it gives, e.g. Government supported developments such as aircraft, and wrong when it demands, especially in the field of information.

'(d) We must be vigilant to assert that personal choice is one of the mainsprings of human endeavours, and that restrictionist intervention, by denying choice, saps initiative and so economic growth.

'(e) Finally, in cases where we are not clear about the ends, and accept the means, or even in cases where we are not clear but Parliament had ordained, we must co-operate in a real sense of partnership.'

Definition of views and attitudes apart, there were four respects in which Government policy was making a deep impact on Vickers affairs in 1965 and 1966. Cancellation of the TSR2 project was announced in April 1965. A Committee of Inquiry into the Aircraft Industry, under the chairmanship of Lord Plowden, reported at the end of 1965. A Shipbuilding Inquiry Committee, with R. M. Geddes as chairman, reported in March 1966. In the following month the Government announced in the Queen's Speech its intention of introducing legislation to nationalise the steel industry.

A profoundly pessimistic view of the TSR2 cancellation was taken by Sir George Edwards and his colleagues at BAC. The project had been at the heart of the military aircraft programme for over six years, and a prototype was flying successfully. To cancel now, despite escalating costs, in order to spend dollars by the billion on an American aircraft of doubtful merit (itself cancelled two years later) seemed both nationally and industrially incomprehensible. BAC in its present form, Sir George Edwards told the Vickers Board, was entirely dependent on the TSR2 contract. In an attempt to prevent cancellation BAC had offered to complete the contract at a price of £430m and to sacrifice all profit if the cost exceeded that figure, but the offer had been rejected. Cancellation would now mean works closures and the loss of up to 20,000 jobs by the end of 1966. For BAC to remain in being as a viable company it was essential that contracts should be obtained for both the proposed Anglo-French collaborative projects—the Jaguar supersonic strike-fighter trainer and a variable geometry combat aircraft.

In the event BAC rolled with the blow remarkably well. Both collaborative contracts were won, not unexpectedly in view of the lead the Corporation had established in supersonic know-how. Then at the end of 1965 orders for Lightnings (the first fully supersonic aircraft to go into service with the RAF) and Jet Provosts came from defence contracts placed in Britain by Saudi Arabia. With the BAC One-Eleven breaking into the international market for civil aircraft, the Concorde project proceeding satisfactorily and the Guided Weapons Division enjoying considerable success with the Vigilant, the TSR2 cancellation came to be seen more as a setback than a disaster, though for the thousands who lost their jobs this might have seemed an excessively philosophical assessment. Financially the immediate effect of the cancellation was to check the slowly improving profitability of BAC, but in the longer run cancellation charges were paid which ensured that no loss was suffered on the project. The most damaging effect in the short term was on confidence, not only within the Corporation, but among potential customers for aircraft such as the One-Eleven. A question mark had been raised over the future of BAC. The fact that it hung there for a relatively short time reflected the energy with which the situation was tackled. 'BAC survived the cancellation because of powerful and determined management in the team up north,' said Sir George Edwards. 'There was sufficient flexibility to put work into Preston to provide a breathing space. There had to be massive reductions and cost trimming but we put in what work we could and then along came the two collaborative projects.'

Within Vickers House the cancellation intensified doubts about Vickers' continuing involvement in the aircraft industry, and these doubts coincided with the taking of evidence by the Plowden Committee which had been set up at the end of 1964. When the Committee reported in December 1965 it recommended (with a reservation by Mr Aubrey Jones) that the Government should acquire a shareholding in BAC and in the airframe elements of Hawker Siddeley, including the guided weapons interests. This, the Committee thought, would provide the basis for 'a suitable partnership between public and private capital in the circumstances peculiar to this industry'. When the Vickers Board discussed the recommendation they came to three principal conclusions. First, that the prospect of securing a reasonable

return on the shareholding in BAC was remote. Second, that BAC
could not prosper unless it continued to develop new aircraft and
modified versions of existing aircraft, and resources were not available
for this. Third, Vickers policy should be to negotiate the most favourable arrangement possible with the Government.

Government policy was set out by the Minister of Aviation (Mr
Fred Mulley) in a debate in the House in November 1966. It was not
only a question of Government participation in the equity of BAC
and Hawker Siddeley, he said. There was also the question of a merger
between the two companies. The Government had always found the
working capital for military projects, but they were now expected
to find the greater part of the risk capital for major civil projects as well.
They had therefore come to the conclusion that 'the national interest
would best be served by a merger of the airframe interests of BAC and
Hawker Siddeley into a single company in whose equity the Government would take a substantial minority interest'. He went on to say
that the two companies had indicated their willingness to co-operate
and that negotiations would shortly begin.

In fact, Vickers could see no advantage in an arrangement which
would leave them in a minority position in a larger grouping. An
alternative was for Government either to acquire BAC or to buy a
minority holding. If they bought the whole of BAC they could then
decide whether to go ahead with the merger. If they bought a minority
holding, and did the same in Hawker Siddeley, they would be in a
strong position to determine global policy since they already had
60 per cent of Short & Harland and owned Beagle. Vickers approach
was further conditioned by the knowledge that, as compared with
Hawker Siddeley, BAC held all the new projects and had greater
experience in the design and building of large and supersonic aircraft.

After an initial spurt the negotiations made only slow progress,
possibly because there had been ministerial changes and the new
Ministers, Mr Benn and Mr Stonehouse, wished to re-think the
Government position. After several months of silence about Government intentions the chairmen of the principal companies wrote a
formal letter to the Minister of Technology in November 1967 to
protest that forward policy making was being hampered and morale
undermined by the Government's delay in declaring its intentions.
This had the desired effect. In mid-December, a Written Answer by
the Minister said that following devaluation the Government was
re-examining its policies 'in a number of fields', and it would not
therefore be possible 'for the time being' to proceed with the negotiations. 'I should like to reiterate, however', the Answer concluded,

'that in the Government's view a merger of the aviation interests of these two companies is desirable.'

Despite earlier doubts about the value of the investment in BAC Vickers were well content with this decision. Recent results from BAC and the state of its order books, said Sir Leslie, showed that the Corporation was 'in an altogether healthier position than could have seemed possible a few years ago . . . Our investment is to be regarded as a good one.'

There was more to come, however. In mid-1968 it was learned that Rolls-Royce, who had acquired the Bristol interest in BAC in October 1966, now wished to dispose of it and had told Hawker Siddeley that they could have first refusal if the other two partners, Vickers and English Electric, chose not to take it up. The Rolls-Royce holding was 20 per cent as against the 40 per cent each held by Vickers and English Electric, and under the founding agreement any partner wishing to withdraw had first to offer his holding to the other two. For Vickers it was inconceivable that 20 per cent of BAC should be acquired by its main competitor, with the insight this would give into its operations, and effectively the new situation meant either that Vickers and English Electric must take 10 per cent each or that Vickers must take the whole 20 per cent.

The plot now thickened, with the announcement of a bid for English Electric by Plessey. If the bid succeeded it was known that Plessey would probably wish to divest themselves of the 40 per cent holding in BAC. The possibility therefore existed that Vickers would be faced with the need to decide on taking up the holding of both the other partners, in other words of resuming the role of aircraft manufacturer in its own right, a role it had deliberately cast aside in 1960.

In this situation the Government saw an opportunity to move towards the single airframe manufacturer which was its objective, and the Industrial Reorganisation Corporation, one of the parastatal organisations of which Sir Leslie Rowan had spoken, accordingly appeared on the scene. The best course, they suggested, was to build a single airframe unit around Hawker Siddeley. Not on your life, said Vickers. BAC had made a remarkable recovery from the TSR2 cancellation and from the uncertainties arising from the Plowden Report: its five-year projections were good and what it now needed was an untroubled period of consolidation: nothing would be more damaging than to revive all the old apprehensions and doubts: and, in any case, the relative positions of BAC and Hawker Siddeley had changed greatly. In short, Vickers told IRC, they believed that there was room in the airframe industry for two main operators, and nothing should be done to disturb the rapidly growing success of BAC.

The possibility was then discussed of putting BAC shares on the market. Here Vickers' representatives expressed the view that on the evidence of results in the immediately preceding years a successful placing would be unlikely, though in another three to five years time the position would probably be very different.

When Vickers Board discussed these developments in September 1968, they showed no enthusiasm for increasing the Company's stake in the aircraft industry, while recognising that this might be necessary. Certainly they felt that Vickers investment would be gravely affected if Hawker Siddeley were to gain a foothold in BAC. Possibly a satisfactory outcome would be for Vickers to have a majority interest, with IRC holding the balance pending a public issue?

At this stage the scenario changed yet again. In November 1968 GEC successfully entered the lists in opposition to Plessey for control of English Electric and a new figure took the stage. This was the redoubtable Arnold Weinstock. With GEC now holding 40 per cent of BAC, his attitude to the holding was of cardinal importance. His interest, it seemed, lay not in the airframe activities of BAC but in its guided weapons. He would happily discard the airframe business, but on the other hand did not wish to rock the boat and for the time being at least would stay with the situation as it was. He did so to the extent of agreeing later to join with Vickers in buying the Rolls-Royce holding, which was formally offered to the other partners when Rolls-Royce went into voluntary liquidation in 1971. Vickers and GEC thus each acquired another 10 per cent to make them joint owners of BAC.

The upshot of the GEC appearance on the scene in 1969 had thus been to provide the period of stability for BAC for which Vickers had so strongly argued. That the argument had been well founded became abundantly apparent in 1973 when BAC produced a dramatic upsurge of profits, an upsurge which was to gain increasing momentum in the following years.

The Plowden Committee's report on the aircraft industry was published in December 1965 (Command 2583), and the Geddes Committee report on the shipbuilding industry (Command 2937) followed four months later.

The Geddes Committee had been given the task of advising on the changes necessary to make the shipbuilding industry competitive in world markets. The main Geddes recommendations were for reorganisation of the industry into groups and for setting up a

Shipbuilding Industry Board 'to stimulate the re-organisation of the industry' and to 'control Government financial assistance for reorganisation'.

In support of geographical groupings the Committee cited four advantages.

(a) A strong management organisation would be most effective if its contact with the yards was close, quick and personal.
(b) Economies of grouping, such as central drawing, planning and purchasing offices, would be facilitated by close proximity.
(c) There could be flexible use of the labour force between yards.
(d) Morale could be built on regional and 'river' loyalty and pride.

The thinking of the report was broadly approved by Government and the recommendation accepted for the setting up of a Shipbuilding Industry Board. With the passing of the necessary legislation the Board came into being in 1966 under the chairmanship of Sir William Swallow. The scene was thus set for the shipbuilding companies to discuss possible groupings.

In fact, even before the Geddes Committee had been set up, Vickers had come to the conclusion that a Tyne consortium would make sense, and discussions with Swan Hunter, Hawthorn Leslie and John Readhead & Sons were already in train when the Committee began sitting early in 1965. The four advantages of a geographical grouping cited in the committee's report were strongly in mind during these discussions, but other factors also existed. One current problem was a shortage of skilled labour, and rationalisation might be a means of using it more efficiently. On a broader view, Vickers could not ignore the poor profit performance of its Newcastle Yard over a long period. In over 40 years of Vickers' operation the yard had losses in 13, and the net gain for the whole period was £4.5m.

There was also the question of whether it was wise to mix naval and commercial shipbuilding. The Geddes Report picked up this point and came strongly to the conclusion that naval shipbuilding should be confined to specified yards. 'A yard which mixes both types of work can be compared with a production line on which lorries and racing cars are intermingled,' it said. 'This leads to inefficiency and is an important source of weakness in the industry.' The report argued for 'the virtual segregation of sophisticated work in specialised yards concentrating on home and overseas naval orders, and such other sophisticated ships as may be required to meet the civil demand'. It noted that this was the position at Vickers' Barrow Yard, 'where no merchant ship is being built'. That Barrow should have become almost exclusively a naval shipbuilder was partly the result of deliberate policy and partly the fact that the nuclear submarine programme had become focused there. At Newcastle the mix of naval and commercial work continued, and Vickers had now to consider whether this was desirable and, if not, what alternative existed. Given the many other reasons in favour of grouping, and given also the poor profitability of Naval Yard, the sensible answer seemed to be in its inclusion in a Tyne consortium. Vickers direct shipbuilding activity would then be concentrated at Barrow, but an investment would be held in the new consortium. It was towards these ends, accordingly, that the Vickers negotiators addressed themselves.

To reconcile the interests of all those involved proved a difficult task, however, and in September 1965 the Vickers Board was told that little progress had been made, beyond general agreement that the shiprepairing establishments (including Palmers Hebburn) should not be included. A scheme put forward by Sir John Hunter, chairman of Swan Hunters, had not proved acceptable to the other companies and it had been agreed to invite Sir Henry Benson to look into the possibility of formulating a scheme. This he duly did early in 1966, recommending a BAC type of organisation, with no parent holding overall control, but again no agreement was achieved and the matter was put into temporary cold storage pending publication of the report of the Geddes Committee. The report did not, in fact, greatly change the situation, beyond holding out the hope that if a Shipbuilding Industry Board came into being some financial lubrication might be available. Discussions were resumed and continued into 1967 when at last agreement was reached in principle and announced in June.

The agreement was completed in December and the new company formally came into existence on 2nd January 1968, under the title of Swan Hunter and Tyne Shipbuilders Limited. Holdings of the share capital were calculated basically on fixed assets which gave the Swan Hunter Group (now including Readheads) 64 per cent, and Vickers and Hawthorn Leslie 18 per cent each. The agreement included arrangements for completing existing contracts on an 'old account' basis. At Naval Yard these arrangements covered HMS *Bacchante*, a frigate for the Royal Navy, and two phosphorus carriers. A frigate for Iran was towed to Barrow for completion. Vickers, as its partners knew, had no intention of retaining its holding in the longer term and two years later, at the end of 1969, the holding was sold to Swan Hunter Shipbuilders for £1.168m, the book value of the assets.

Although, on the issue of groupings within the industry, the Tyne was the focus of interest for Vickers, the company came also under some pressure from the Government through the Shipbuilding Industry Board to bring Barrow into some form of North West grouping. At first a grouping of Barrow, Cammell Laird and Harland & Wolff was proposed, and the possibility of linking Barrow and Harland & Wolff was also examined. Vickers found little that was attractive in these proposals. Barrow was firmly committed to the nuclear submarine programme, it had a satisfactory volume of other naval work, including export orders, it operated profitably as it had always done, and it was both isolated and self-contained, so that none of the advantages seen by the Geddes Committee in proximity to other yards was to be gained from grouping. As to local loyalty and pride, Barrovians already had this in full measure and wanted no interlopers. With enthusiasm also lacking at Birkenhead and Belfast the proposals stood little chance of acceptance and discussions languished. In September 1968 the Government tried a little arm-twisting when Sir Leslie, Mr Yapp and Mr Redshaw were invited to discuss with Mr Benn the future of Barrow should a hiatus occur in defence orders. Would it not be sensible, they were asked, to envisage a merger bringing in not only Cammell Laird and Harland & Wolff, but also Vosper-Thornycroft? Sir Leslie again set out the case for believing that Barrow should and could stand alone, but said that if any merger took place at all it should be confined to Barrow and Harland & Wolff, with Barrow concentrating on sophisticated building and Harland & Wolff on routine work. A month later, at a meeting with the Shipbuilding Industry Board, representatives of all four companies joined in providing evidence to show that a merger of this kind had little or nothing to commend it. To all intents and purposes, that was the end of the matter.

Looking back on the Geddes Report, one of Vickers top shipbuilding managers said some years later that it could have succeeded in revitalising the industry only if all three of its main recommendations had been acted upon – an enthusiastic work force, sustained government support, and companies prepared to sink industrial differences. Of these three requisites only the third had happened. Geddes had also recommended that the industry should be allowed to obtain steel at preferential prices. In fact, steel prices had rocketed.

The computer industry, no less than the shipbuilding industry, came prominently into the Whitehall arena during 1967 and 1968. With ICT the biggest British computer company, Vickers' interests were again strongly involved through its 23.6 per cent holding. When ICT was formed in 1958 by the merger between Powers-Samas and The British Tabulating Machine Company Limited Vickers had 38 per cent of the equity. In 1961 ICT made an issue to existing shareholders with the result that Vickers now had 4,750,000 shares. A holding of this size, the Directors considered, absorbed 'a disproportionately large part of Group funds'. Accordingly 1,375,000 shares were sold (at a profit of £2.545m) and Vickers holding then stood at 27 per cent. Another change took place in 1963 as the result of a further issue of ordinary shares, which added 843,750 shares to Vickers holding at a cost of £2.637m and established Vickers proportion at 23.6. This, then, was the position when talks began in 1967 about a further merger in order to establish a British company capable of holding its own against the American giants.

Speculation that the Government proposed to take a substantial stake in ICT began to appear in the press towards the end of 1967, and at a press conference in December Colonel Maxwell, as chairman of ICT, confirmed that talks were taking place. The participants, he said, were ICT, the Ministry of Technology and English Electric, and they were 'seeking a strong, independent British computer industry by merging the various computer interests'. As the talks proceeded Plessey also became a participant. Vickers approved the broad objective, which it thought to be in the national interest, but had to protect the value of the investment and did not relish the prospect of being locked into a larger company with a smaller percentage holding. Provided that this was understood, and that no objection would be raised to sale of Vickers holding at a later stage, Vickers were prepared to co-operate in pressing the talks to a satisfactory conclusion. This was achieved in March 1968 and Vickers holding in the new company, International Computers (Holdings) Limited, represented 12.6 per cent of the equity, though with the same book value of £9.336m.

If a positive role had been played by Government in the computer merger, its role in relation to orders for Chieftain tanks seemed to Vickers to be somewhat less than straightforward. Vickers were design parents of the Chieftain and had received one-third of the first order for the Chieftains required for the British Army. The other two-thirds went to the Royal Ordnance Factory at Leeds, with the advantage that gave them from a longer production run. To them also went the order for spares. In an announcement in March 1968 about the second order the Government said the whole order would go to the ROF, and this conformed with Labour policy of trying to concentrate production in the Government factories. For Armament Division at Elswick the announcement was a serious blow for once it had completed work on the first Chieftain order it could survive only on export orders. The prospect therefore existed of many hundred redundancies, not to

9

mention the ending of Vickers tank design capacity. Consternation in Newcastle spilled over into the national press and the Government found itself under pressure to justify the decision. Mr Roy Mason, then Minister for Defence (Equipment), said that Vickers had been told that their first tender was too high and had been given an opportunity to tender a second time; they had been given 'two bites at the cherry', but had failed to respond. This line was echoed by two Labour MPs in Newcastle, Mr Edward Short (then Postmaster General) and Mr Robert Brown. According to press reports, Vickers prices were said by them to be 10 per cent higher than those of the ROF.

Behind these allegations was a history of dispute about the make-up of costs by Vickers and by the ROF. The Army Department had said that Vickers prices compared unfavourably, and Vickers had responded by saying, in effect: 'Well, let's be sure that like is being compared with like' and proposed a joint examination of costings. This invitation the Department declined. When the public 'row' flared up in 1967 Vickers protested to the Minister, invited the two Newcastle MPs to come and discuss the matter, and again proposed an examination of costings, this time by an independent firm of accountants. Again the proposal was rejected, understandably enough since in Vickers knowledge the ROF costs carried neither design overhead nor overhead for headquarters administration. The hard fact remained that no further order had been placed with Vickers, and since no immediate redundancies were involved, and since Vickers roundly declared its intention to maintain the division as long as it could, by export orders, the issue faded from the headlines. In August it reappeared briefly when a report by the Public Accounts Committee included a section on the Chieftains. In this section the Department's allegations about Vickers were regurgitated though Vickers had neither been told that the matter was coming before the committee nor been invited to give evidence.

In the event, and in the nick of time, Vickers obtained export orders, not only for its own battle tank but also for other equipment, and Armament Division remained in business. In the national interest this was just as well because the capacity at Newcastle became of prime importance when the Government negotiated a major contract for Chieftain tanks for Iran a few years later.

5

Exit Steel 1965–68

1967 was the year in which Vickers celebrated its centenary as a public company. It was also the year in which Vickers finally lost the steel business in which it had been founded in the 1820s. 'Lost' is perhaps as good a euphemism as any for a forced sale to the State. The British political system being what it is (and Winston Churchill, while recognising its failings, could still think of none better), there could be no doubt that after the 1966 General Election, which increased its majority to 97, the Labour Government would succeed in carrying legislation to nationalise the steel industry. Whether nationalisation truly reflected the 'will of the people' might be a nice debating point, but the reality of the situation was plain enough and Vickers had mainly to be concerned with obtaining the fairest possible price.

The likelihood that a new Labour Government would re-nationalise the steel industry had always been on the cards. Because of this likelihood the decision to buy back the English Steel Corporation in 1954 had been taken by the Board with a certain reluctance. Early in the Knollys' chairmanship studies of the situation were undertaken with a view to seeing whether the Corporation could be reorganised in such a way that it would fall outside any new Act, either completely or partially. Papers were prepared by Frederick Pickworth, Chairman of ESC, and by Eric Faulkner, and advice was taken from Sir Sam Brown.

In fact, the main card had already been played unsuccessfully before the 1951 Act. This was to argue that ESC did not really constitute part of the iron and steel industry, but was a business set up as an offshoot of the engineering industry in order to meet the requirements of that industry for specially treated products. When this argument was rejected in 1949 attention had been turned to obtaining exclusion of those parts of the business not falling clearly within the activities specified in the legislation. The Government showed a disposition to accept that a case existed for some activities remaining with private industry, but would agree only to discussion of these possibilities between Vickers and the Iron & Steel Corporation after the whole of ESC had been nationalised. In the event the Corporation declared their readiness to 'hive off' the Engineers Tool Division at the Open-shaw Works and the File Factory at Holme Lane, Sheffield. Vickers and Cammell Laird accepted this offer in principle, but before terms could be negotiated the Government fell and the new Conservative Government proceeded to de-nationalise the industry.

Against this background it could not be expected that a new Labour administration would listen sympathetically to arguments on the same lines. On the contrary, all the indications pointed to a thoroughly inflexible attitude. The next Labour Government, said Aneurin Bevan, would re-nationalise the industry in such a fashion that no subsequent Parliament would be able to disturb the arrangement. Moreover, it was declared from Labour Party sources that the company structure would be broken up and much stricter political control exercised over the successor to the Iron & Steel Corporation.

Very few alternative lines of action suggested themselves as a result of the studies within Vickers. One was to transfer the shares in ESC to an overseas company, such as Canadian Vickers, but this was quickly seen to be an idea more ingenious than practical. Total exclusion, it was felt, would be impossible. What of partial exclusion? For adminis-trative purposes ESC had already been reorganised as a holding company with seven manufacturing subsidiaries in the UK, each with its own group of products. Perhaps this would facilitate the case for hiving some off as more engineering than iron and steel activities? Here again, closer examination with the aid of Sir Sam Brown pro-duced gloomy conclusions. Those subsidiaries for which the strongest case could be made for exclusion happened also to be those showing least profitability.

Given these conclusions it was perhaps remarkable that a decision should have been taken to proceed with the Tinsley Park project at such great cost. There could be no assurance that the magnitude of this investment — some £26m — would be adequately recognised by a new Labour Government if and when terms had to be negotiated. However, the robust view prevailed that action should proceed as though no threat existed. By 1965 the profits of ESC should be enhanced by earnings from Tinsley Park, and Vickers interests apart, the project must be to the benefit of the steel industry and the country.

Tinsley Park was, in fact, nearing completion as the second Mac-millan administration approached the end of its term of office. A Labour victory in the next election had clearly to be envisaged, and early in 1963 Vickers set up an internal working party, headed by J. H. Robbie, to consider with colleagues from ESC 'whether any institutional arrangements can be made to exclude from a possible nationalisation of ESC any parts thereof which the parent companies would wish to retain'. When the working party reported in September, however, they could do little more than go over the old ground, set out the various possibilities that had been considered and the reasons why none seemed acceptable. The Board discussed the working party's report in October and decided that the matter should be allowed to rest pending further developments, if any.

The situation became 'live' again towards the end of 1964 when the Labour Party won a narrow victory at the general election. Re-nationalisation of the steel industry figured prominently in their manifesto and in anticipation of legislation Vickers, Cammell Laird and ESC set up a committee to keep the matter under review and to advise on action. Membership of the committee included Sir Henry Benson, a chartered accountant now at the top of his profession, with a reputation for wisdom and skill in financial negotiations that made him much in demand as an adviser. In fact, there was no immediate introduction of legislation and events marked time until the Labour Party had obtained its decisive majority in the election of March 1966. The intention to legislate was then declared in the Queen's Speech, and the battle was joined.

It was, of course, a battle involving the interests of many companies besides Vickers and Cammell Laird, and an attempt had accordingly been made to arrive at a common view. In May 1965 Sir Charles Dunphie was among the signatories of a statement by the chairmen of the major steel companies. The statement agreed that the efficiency of the industry would be enhanced by measures of rationalisation, but asserted that rationalisation could best be achieved without the disadvantages of public ownership. To produce plans for rationalisation the Iron & Steel Federation accordingly set up a development co-ordinating committee under an independent chairman.

In the meantime, as was noted in Vickers annual report, 'the uncertainties surrounding the future of the steel industry persisted into 1966 and inhibited policy making'. One possibility thus put into cold storage was a merger in some form between ESC and Firth Brown.

What had become increasingly clear was that the Government was in no mood to listen to representations that would moderate in any way its intention to obtain nationalisation. The arguments for rationalisation without State participation were brushed aside, and Vickers had to accept that ESC would be swallowed whole. It was now simply a matter of negotiating the best possible terms, and Sir Henry Benson was appointed by Vickers and Cammell Laird to represent their interests in the negotiations with the Minister of Power.

With ESC a subsidiary company, and not therefore quoted on the Stock Exchange, the price to be paid had to be negotiated within the terms of a formula presented in the Act. In effect, this meant estimating the price at which the ordinary shares would have stood on specified dates had they been given a Stock Exchange quotation.

In making this calculation, comparison with quoted companies was a factor of major relevance. To arrive at precise comparisons was obviously difficult, but three were selected – United Steel, Colvilles and Hadfields. Other 'relevant factors' in relation to ESC were identified as the financial and management backing of the parent companies, the element of tied business with the parent companies, the wide spread of interest in steel and engineering industries and the relative stability of the Corporation's record in comparison with many quoted steel companies. Taking all these factors into account, Sir Henry Benson put forward the view that a fair valuation for each £1 of paid-up ordinary share capital in ESC was 42/-, equivalent to £25,725,000 for the whole of the ordinary capital. For Vickers, as holders of 75 per cent of the ordinary shares, this would represent compensation in the order of £19.3m.

Though vesting date was 28th July 1967, the compensation negotiations took a further six months. The figure then agreed amounted to £16,345,328 in respect of Vickers holding. Payment was made in the form of Treasury Stock on 29th January 1968, and this stock was sold by Vickers on the following day for £16,332,000.

Arguably the price would have been better had Tinsley Park been able to show results in 1964 and 1965 nearer to those forecast. The plant had become operational by stages during 1963 and it had been assumed that operational losses would be made during the running in period in 1963 and 1964. These had been estimated at £790,000: in fact, they totalled over £1m. In 1965 it had been estimated that there would be an operating profit of just over £2m: the actual result was a loss of £574,000. The main reasons for this disappointing performance lay in the generally depressed state of industry, which meant that orders failed to give full loading, and a rapid escalation in wage, fuel and other costs which could not be adequately covered in increased prices. To a large extent, it could be argued, these conditions arose from Government measures to dampen down the economy, but the fact remained that on figures Sir Henry Benson did not hold a strong hand. Operational losses at Tinsley Park, added to heavy interest charges arising from borrowings to finance the project, meant that

ESC profit before tax showed a sharp decline from £4.1m in 1961 to £1.4m in 1965. The compensation negotiations thus took place at an exceptionally low ebb in the company's fortunes.

There could be no doubt, however, that for £16.3m the State had acquired a massive asset. 'The ESC group,' said Sir Charles, 'supplies a wider range of steel and engineering products than any other steel company or group in the United Kingdom: it is equipped to produce larger and heavier ingots, forgings and castings than any other British steelmaker: it is the major British producer of special steels: and in the new works at Tinsley Park it has Europe's most modern alloy steel plant.'

For Vickers part, the arithmetic showed a net loss of just over £4m. Net assets of ESC attributable to Vickers at the end of 1966 totalled £20.4m against compensation of £16.3m. Not calculable was the loss of one of the four main pillars on which the Company had been built.

Whatever view might be taken about the fairness of the settlement, Vickers now had to decide how to use the money. Three main choices presented themselves. It could be paid out to stockholders: it could be used to promote growth, whether in existing activities or by acquisition of new businesses: or it could be used to pay off indebtedness.

The case for payment to stockholders did not look very strong and the Company was not put under pressure to take this course. In a company such as Vickers, with a wide spread of activities, disposals and acquisitions occurred with some frequency, and funds from disposals helped to finance growth. The re-nationalisation of ESC represented an enforced disposal, and the amount was large, but the principle remained the same. The Company had to be aware, however, that in retaining the funds it must demonstrate that they were being used to the long-term advantage of the business and therefore in the long-term interests of the stockholders. Apart from these considerations, there were practical objections to distribution since it would almost certainly involve stockholders in tax liabilities, both in income tax and capital gains tax. Moreover, if the distribution were in the form of capital the sanction of the Court would be necessary and this might well be refused with company borrowings at a high level.

In debating the relative merits of investment in growth and paying off indebtedness, the Board had to recognise that for some years the Company's difficulties had stemmed largely from the burden of indebtedness, brought about particularly by the VC10 project. Interest charges were seriously depleting profits, and this at a time when profits were being eroded in other ways. Low profits meant a chain of other problems, to the extent indeed of putting the Company at risk. Paying off indebtedness would both win time and establish a new base from which to expand later. With these factors in mind, the Directors instructed that the proceeds should be applied to the reduction of bank loans, the FCI loan and the acceptance credits, and that the balance should be invested in short-term loans. The effect could be seen in comparison of loan and interest figures for 1967 and 1968. Short-term loans at the end of 1968 stood at £6.5m compared with £14.5m at the end of 1967, and interest payable during 1968 totalled £1.14m against £2.23m in 1967, an improvement of just over £1m. However, these figures had to be seen in conjunction with the loss of some £1m in investment income from ESC, so that the immediate net effect of re-nationalisation on Vickers 1967 profit and loss account was insignificant.

6

Prone to Accident 1965–70

While Vickers wrestled simultaneously with the problems of re-organisation in the aircraft, steel and shipbuilding industries, much else was also happening, most of it worrying.

In the mid-sixties the Company's profit before tax had fallen to levels well below half that earned a decade earlier. As then, however, a 10 per cent dividend was still being paid. The Knollys dictum about covering a dividend at least twice no longer held. In 1966, 1967 and 1968 the dividend was not covered and in 1969 only just covered. Arguments for and against a cut were finely balanced. In 1969 the no-cut school felt that at a time when the Company was beset by troubles it would be a very serious blow to confidence, both externally and internally, for Vickers not to pay its 'traditional' 10 per cent: a good deal of undistributed profit had been accumulated from previous years: it would be unfair to stockholders, who had received no part of the steel compensation money: and in any case projections for the future were promising, especially as the 'new' activities gathered strength and momentum and as risk areas were reduced by closure and disposal. The case for not cutting was reinforced, moreover, by City advice which thought that a cut would have a major adverse effect on the shares and that this in turn would open the way for a take-over. Arguments for making a cut were based on the impropricty in financial orthodoxy of paying out in successive years more than had been earned (in effect, distributing capital): the need for funds to finance growth: and the belief that the City would see in recent setbacks sufficient justification for a cut without necessarily taking a pessimistic view of the future.

Much of the difficulty in arriving at a firm view lay in the uncertainty engendered by repeated failure of forecasts. Not only were many wrong forecasts coming forward from operating units but they were being accepted at the centre and incorporated in management plans. Moreover, each year seemed to produce at least one unforeseen 'disaster'. To make matters worse, neither adverse trends nor specific situations were being detected at the centre in time for remedial action to be put in hand. It seemed clear that, among other things, some strengthening was necessary in the financial organisation at the centre so that there could be both more effective financial evaluation and more effective monitoring. To this end J. S. Booton, FCA, joined the Company as Comptroller in June 1967, and Malcolm Paris was appointed Assistant Comptroller a few months later to carry out 'specific non-routine tasks'. Both were young men, already identified as potential 'fliers'. Booton's immediately preceding appointment was as Group Chief Accountant of Viyella International and previously he had held a similar appointment with Powell Duffryn. Paris also came to Vickers from Viyella.

These appointments represented part of a second phase of reorganisation associated with Sir Leslie Rowan's succession to the chairmanship at the 100th annual general meeting in June. He had formally relinquished the responsibilities of Managing Director a month earlier and these responsibilities were now assumed jointly by R. P. H. Yapp and W. D. Opher. To Mr Yapp fell particular responsibility for group strategy, deployment of resources and major policy matters in relation to contracts, together with oversight of the activities of the shipbuilding group, the medical group and Ioco. Mr Opher took on oversight of the engineering, printing machinery and office equipment groups and of Vickers Instruments. To succeed W. D. Opher as chief executive of the engineering group J. H. Robbie relinquished his appointment as Director of Finance and this was assumed by A. M. Simmers, hitherto the Company Secretary. Two new appointments were made to the Board. Leonard Redshaw succeeded George Houlden (on retirement) as chief executive of the shipbuilding group, and A. P. Wickens became director in charge of overseas interests. The successor to A. M. Simmers as Secretary was H. E. Scrope, who had joined the Vickers House staff from the RAF in 1947 and had been Secretary of Vickers-Armstrongs since 1959: he had a deep knowledge of aircraft in general, and Vickers aircraft in particular, and was himself a qualified pilot with experience in air races, including the King's Cup. When in 1958 Vickers completed 50 years of aircraft design and manufacture he produced an authoritative study of Vickers aircraft during this period under the title *Golden Wings*, a privately circulated publication.

If problems were abundantly apparent in 1967, it could at least be claimed that under Sir Charles Dunphie's chairmanship the Company had come through a difficult period of transition with a good deal to count to credit. The financial incubus of the VC10 had at last been removed, BAC remained intact with promise for the future, steel had been taken away, with a big loss in earnings, but acquisitions had been made in growth activities, changes in defence policy had still left Vickers with a central role in the manufacture of arms equipment, particularly for the Royal Navy, new planning machinery and methods had been introduced with McKinsey aid and management training and development procedures had been further strengthened. In his valedictory address, Sir Charles said that he had 'greatly valued the privilege of being a Vickers man'. As a non-executive director, he was to continue as a Vickers man until 1970 so that his service to the Company spanned a period of 22 years, of which six years had been as Managing Director and five as Chairman.

In a brief respite from the problems immediately to be faced, Sir Leslie presided in the summer of 1967 over the celebrations to mark Vickers centenary as a public company. Ceremonies and functions of various kinds were arranged throughout the group, with a contribution of 10s per head from Head Office towards the costs. For some time there had been discussion about making an award to long-service employees, and centenary year marked the introduction of a scheme to present gold wrist watches to all who had completed 40 years service with the UK companies.* Forty years might not seem an unduly generous benchmark, but in Vickers long service was commonplace. 2,144 employees qualified for the watch. Recording the longest service of all, a few weeks short of 56 years, was A. E. R. Bloomfield, a setter-operator in Boby's, beating by a short head another Boby man, F. O. Johnson, a joiner. In London the centenary was celebrated by a month-long exhibition at the Design Centre, where the visitors were headed by HRH the Duke of Edinburgh, and by the production of a film 'Centenary' by the Vickers House film and photographic department which had been set up as long ago as 1916, one of the first units of its kind in British industry. The department was now headed by Leslie Sansom, FRPS, a man much admired and liked in his profession, and 'Centenary' was produced under his direction. It won a good deal of acclaim, including a Gold Award from the British Industrial & Scientific Film Association. Setting Vickers own story in the context of historical and social change, it vividly recalled the Company's contribution to British survival in two world wars and to industrial evolution during the past century – a timely reminder, perhaps, of high achievement and difficulties surmounted.

Certainly, a strong nerve was now needed. The Company, it seemed, had become accident prone. Any hope of plain sailing after

*Reduced to 30 years from the beginning of 1978.

the VC10 provision had been shortlived. With cancellation of the TSR2 the whole future of BAC had for a time hung in the balance. In 1965 the first provision, for £3.1m gross, had to be made against the A. H. Salt contract. In 1966 the troubles besetting Naval Yard's contract to build five cargo-liners for Alfred Holt & Co came to a head and a provision had to be made of £3.4m gross. In 1967 not only was a further provision necessary for the A. H. Salt plant, this time for £2.2m gross to make a total of £5.3m, but also Canadian Vickers had a disastrous year, with a pre-tax loss of £2.2m. 1968 produced three serious setbacks—a prolonged strike at Barrow, a heavy loss by the Hydraulics Division and freak floods at Crayford and Dartford Works. 1969 saw the destruction in two fires of a major part of the Romford Works of the Office Equipment Group. In 1970 provisions had to be made against two Vickers-Zimmer contracts in the German Democratic Republic, totalling £1.55m, and also a further provision against the Riyadh hospital contract in Saudi Arabia which brought the Medical Engineering loss in that year to £1.2m.

The Holt contract had been taken at Naval Yard with something of a flourish of trumpets in 1964. 'The largest single order the Yard has received for cargo liners,' it was announced. 'We are very glad to continue an association with this great shipping company for which we have built sixteen ships since 1947'. In 1964, however, building orders were a matter of intense competition. Too many contracts, the Geddes Report noted, 'were at prices so low as to give too small a profit or even a loss'. So it was with the five *Priam* ships. The tender had included full charges but no profit. With a run of five vessels of the same class, Naval Yard hoped it could reduce costs per vessel as the work proceeded and so emerge on the right side at the end of the contract. This was a forlorn hope. The ships were of sophisticated design, including a considerable degree of automation. Not only was the work more difficult and complex than expected, but the time scale for completion was unrealistically estimated. Nor did variations in the owners' requirements during building help matters. So far from costs being reduced they escalated, particularly on wages. Moreover, shortages of labour in key trades contributed to increasingly serious delays, To make matters worse, divergencies of view existed within the Shipbuilding Group about management handling of the contract. At Naval Yard it was felt that by employing consultants access would be ensured to the most up-to-date and sophisticated management procedures and techniques. At Barrow the view was taken that this could lead only to management by paper and committee and to loss of contact on the berths. Naval Yard duly employed consultants – encouraged, thought Barrow, by Vickers House, though this contention would certainly not go unchallenged – so that when things began to go wrong Barrow felt able to say: 'I told you so.' In fact, the precise degree to which particular factors contributed to delays and escalating costs could only be a matter of opinion, especially since the Japanese builders of two other *Priam* ships were also reported to have suffered substantial losses.

Whatever the niceties of the matter, Holts inevitably felt alarmed and aggrieved. As a result of representation at chairman level, the decision was taken to put Naval Yard in charge of William Richardson, then General Manager of the Barrow Yard. Illness prevented his taking up this new responsibility immediately but another senior Barrow man, William Parnell, travelled hot foot across the Pennines to hold the fort for six weeks. 'Bill' Richardson enjoyed a considerable reputation for the successful management of major projects, notably *Oriana* and the nuclear submarine programme, and was admired for the way in which he combined friendly patience in discussion with firmness in decision. His thorough understanding of the shipbuilding business was apparent in papers read before professional bodies, including papers on merchant shipbuilding and warship building in the UK and on the use of steel by the shipbuilding industry. On arrival in Newcastle he succeeded to an important extent in retrieving the position, partly by arranging for the use of Blyth Shipyard which had been closed in 1967. Blyth had available not only a 40-ton crane but also space to bring a ship alongside. By use of Blyth it was possible to outfit two ships at the same time, and the additional cost of towage to Blyth, and transporting men there, was justified by the urgent need to catch up on schedule. This was achieved to the extent that the contract was completed in November 1967, some $10\frac{1}{2}$ months late.

Technically the five ships proved excellent examples of shipbuilding skills, but the combination of delay and escalating costs left Vickers smarting under a pre-tax loss of £3.4m. After tax the figure amounted to £2.04m of which £2m was debited to contingencies reserve. This experience played its part in the decisions to put Naval Yard into the Tyne merger and to concentrate Vickers shipbuilding at Barrow in the role which the Geddes Report later designated as an S yard, engaged in naval shipbuilding and such merchant ship work as might call for a high proportion of fitting-out capacity.

Following the Holt provision in 1966 came the heavy loss by Canadian Vickers in 1967, mainly in the Shipbuilding Division. When Vickers was contemplating the re-acquisition of Canadian Vickers in the early 1950s they had despatched Peter Muirhead to

Montreal to report on the business. In general his report was favourable, but he had commented: 'Shipbuilding in Canada is bound to be a fluctuating and hazardous business.' It had taken some time for this chicken to come home to roost, but now it had happened. The total loss was £2.283m, mainly on shipbuilding (as distinct from ship-repairing). In reporting the loss, Sir Leslie said in the 1967 annual report: 'The shipbuilding losses were due in large measure to wage escalation, which cannot be recovered under the terms of the contracts, and were accentuated by the effect of an eight-week strike in the middle of the year, which was followed by the negotiation of a new wage agreement. There were also shortcomings in management and production. Steps have been taken to strengthen the management and to improve production methods.' The loss was shared to the extent that 46.25 per cent was borne by the minority shareholders. The net loss attributable to Vickers Limited was £1.183m. With this demonstration that shipbuilding in Canada was indeed a fluctuating and hazardous occupation, the Board of Canadian Vickers recom-mended that the company's activities should be concentrated on activities on the industrial division and on shiprepairing. Shipbuilding was accordingly phased out. The processes of doing so included sale of the George T. Davie Yard at Quebec and were completed in September 1969. One of the last ships to be delivered from the Montreal yard was the *Louis S. St Laurent*, the world's most powerful icebreaker using non-nuclear propulsion. The withdrawal may have been inevit-able, but it involved a good deal of regret. As the annual report noted: 'The decision to discontinue shipbuilding necessitated the demolition of the covered shipbuilding ways from which, over 58 years, have been launched the lead ships for every class for the Canadian Navy completed during that period, as well as hundreds of ships for fresh and salt water service.' That the decision was justified, however, could be seen in subsequent results from the company.

In the 1968 packet of misfortunes the most serious items were a prolonged strike at Barrow, beginning in July and continuing into 1969, and a loss of over £750,000 by the Hydraulics Division. Calcula-tion of the loss inflicted by the Barrow troubles could be made in various ways, but in his report to the Board on the 1968 accounts J. S. Booton assessed the reduction in trading profit at £1.168m. By comparison the financial setback of the flooding at Crayford and Dartford Works in September 1965 was not great, though still assessed at a reduction in trading profit of £199,000. Another setback, estimated at £169,000, came from Palmers Hebburn where dockings for repair had fallen sharply because of a threatened national strike by engineers.

The total reduction in trading profits from these causes during the year thus reached £1.536m. With the loss from Hydraulics to be added, the 1968 results fell well below the management plan. Even so, profit before tax amounted to £7m, which was appreciably better than the 1967 figure of £4.7m.

The extent to which failings or omissions by management could be held responsible for these 'accidents' had to be a matter of separate assessment. The overall situation in the shiprepair industry certainly lay beyond the control of Palmers Hebburn, and it could hardly be said that the floods in Kent were anything but an act of God. Both at Crayford and at Dartford the men in charge, led by A. W. Taylor and L. G. Gooch respectively, minimised the setback by the speed, energy and efficiency with which they organised 'Operation Dry-out', and in this they had splendid support from the staff and work force. At Crayford flood water from the River Cray had swirled through some of the shops to a height of four feet, and when water from the River Darent poured across the Dartford site it had been necessary for everyone to be evacuated. In financial terms, Crayford and the Engineering Group suffered more than Dartford and the Office Equipment Group.

Judgment on the industrial relations dispute at Barrow was eventually rendered by a Court of Inquiry under Sir Jack Scamp. The Court was appointed by the Secretary of State for Employment and Productivity on the 18th February 1969, sat in public at Barrow Town Hall on the 25th and 26th February, visited the site on the 27th, and had its report published as a White Paper on the 28th March.

Attention to the generally unsatisfactory state of industrial relations in the shipbuilding industry had been drawn by the Geddes Report in 1966. 'We cannot emphasise too strongly that improvement in this field is one of the prime conditions of the survival of shipbuilding as a live industry' the report had said. Not least the multiplicity of unions in the industry was noted and the number of disputes arising from demarcation issues. Could not the number of unions be reduced to five, asked the report.

It was precisely this situation that the Barrow management faced. In all, they had to deal with 17 unions and lack of flexibility in agreeing on 'who does what' was seen as the major stumbling block to improved productivity. From time to time agreement would be reached on particular points, but negotiations were usually protracted and progress overall painfully slow. Against this background the incident occurred in July 1968 which led to a strike of fitters, lasting eight months and causing severe delays in production schedules.

The location was the pipe production shop in the shipyard and the trouble started when the management put two plumbers on pipe testing and buffing in the place of two fitters who were transferred to other work. This accorded with an informal agreement made with the plumbers in December 1965, providing for relaxation of pipe testing work to fitters on the understanding that if redundancy occurred among plumbers the work would revert to them. The agreement was not, however, formally notified to the fitters union (the Amalgamated Union of Engineering and Foundry Workers). 'This informality,' the Scamp report noted, 'was a deliberate act of policy and not without precedent, as was subsequently demonstrated. One disadvantage of formal relaxation agreements was said to be that they took roughly three months to negotiate by which time the need for them had passed.' As the two plumbers started work the three fitters still allocated to pipe testing stopped work, as did the other 42 fitters in the pipe production shop. 'It was a commonly held belief among the plumbers', said the report, 'that the AEF District Committee seized on the demarcation issue as an opportunity to involve fitters in strike action in support of their protracted dispute with the Company about apprentices' rate of pay.' The plumbers' union made it clear that if the Company accepted the claim of the fitters to have exclusive rights to testing and buffing work in the pipe production shop then their own labour would be withdrawn. The Company, in short, was 'faced with the virtual certainty of strike action by one trade or the other. The plumbers would strike if the informal relaxation agreement was not honoured: the fitters would strike if it was'.

In its conclusions and recommendation, the Court said they were satisfied on three points:

(a) Over many years pipes had been tested by each of the trades concerned, plumbers, coppersmiths and fitters.
(b) An informal relaxation agreement was made between the Company and the plumbers in December 1965 for some pipe testing to be relaxed to fitters.
(c) The AEF were not notified of this informal agreement.

The Court went on to say that when the AEF first contested the validity of the informal agreement the Company should have accepted that the best course of action was to refer the problem to the Demarcation Court. When eventually the Company agreed to treat the matter as a demarcation issue, use of the procedure was 'frustrated by the continual changes of front by the AEF District Committee'.

Given that longer term policy should be an integrated labour force

Pisces III – the submersible rescued from a record depth after a mishap off southern Ireland.

Vickers in London.

Vickers in Sydney.

Dreadnought, Britain's first nuclear-powered submarine, launched at Barrow by Her Majesty The Queen on Trafalgar Day 1960. Reproduction of painting by Terence Cuneo.

A nuclear reactor, manufactured by Canadian Vickers and weighing nearly 1,000 tons, ready to begin its 1,000-mile journey by barge from **Montreal** to Lake Huron.

ESC ingot.

Her Majesty The Queen at Barrow after launching HMS *Sheffield*, the Royal Navy's first Type 42 destroyer. Lord Robens, Vickers Chairman, on the left and Sir Leonard Redshaw, chairman of the Shipbuilding Group, on the right.

R. O. Taylor, chief executive of Howson-Algraphy, aided by the Lord Lieutenant and Sir Peter Matthews cuts celebratory cake to mark winning of Queen's Award to Industry.

Lord Robens and Sir Peter Matthews in a Vickers communications aircraft en route to a group management conference at Barrow. ▼

in the pipe production shop, the Court concluded by recommending that in the meantime specific rules should be applied whereby fitters alone undertook some of the work with other work shared equally between fitters and plumbers or coppersmiths.

As a final comment the Court said that they were appalled that a dispute concerning the work of 4-6 men should cause such widespread disruption. 'We consider that relations between the management and the unions concerned as a result of this dispute are at a low ebb and that both sides will have to work hard to rebuild confidence. To this end we believe that the Company would be well advised to reconsider their present arrangements for dealing with personnel and industrial relations matters and to ensure that the undivided and expert attention of a senior member of the management team is devoted to such matters.'

The strike and the conclusions of the Inquiry made headlines in the way that industrial troubles always do and industrial achievements do not. It was an episode which caused considerable distress in the Company. Though the financial setback was serious enough the real damage was felt to lie in the damage done to the Company's reputation for good industrial relations. Barrow and the shipbuilders were only part of Vickers and statistics showed that the Group as a whole rarely experienced major disputes. Industrial relations in shipbuilding generally were not good, as the Geddes Report had noted, and to that extent Vickers shipbuilders shared the problems of the industry, with perhaps some additional problems of their own arising from the particular circumstances of Barrow as an isolated community and virtually a company town. Nevertheless, it could be claimed that until the 1968 strike Barrow had experienced no major stoppage for many years: indeed its record in the nuclear submarine programme, particularly in the building of the two Polaris submarines, provided ample evidence of this. As for the rest of Vickers, the annual average of working days lost per 1,000 workers during the five years 1964–68 was 15, less than one-seventh of the average for the United Kingdom.

However, it was no use crying over spilt milk, and Vickers immediately issued a statement describing the report as 'valuable and constructive' and accepting the two main recommendations. The unions also agreed to accept the rules proposed for resumption of work in the pipe shop and so the strike came to an end. If the Inquiry had served to produce a formula to end the strike it had not succeeded in resolving the basic demarcation issues in the pipe-fitting shop. Outside the narrow margins of the agreement fitters and boilermakers remained in contention about who did what, and as a result the shop continued

10

to work at less than full efficiency. This was still the position in 1977.

During the strike no more than one-sixth of the total work force became directly involved, and the ship programmes were remarkably well maintained. Nevertheless, the absence of the fitters caused progressive disruption and demonstrated one of the hard facts of modern industrial life – the power of a relatively small group of workers to bring a major enterprise to the edge of disaster.

'It is a truism,' said Sir Leslie in the 1968 annual report, 'that an organisation's most important asset is the people who work together in it, whether in the offices or on the shop floor. Neither can succeed without the other. Each can be destroyed (in which case both are destroyed) by the hostility or inefficiency of the other . . . Goodwill and good sense on a reciprocal basis are what we are seeking to promote in the new arrangements at Barrow.'

The main feature of the new arrangements on the Company side was the appointment of a director on the Shipbuilding Group Board to take responsibility for industrial relations. The appointment went to J. A. Killick, at that time Deputy General Manager of Barrow Engineering Works. He had served as an apprentice with Vickers in Newcastle in the thirties and apart from six years with the Royal Air Force had spent his career with the Company, his appointments including that of General Manager of Crayford Works. Channels for dealing with industrial relations issues at Barrow had always existed, of course, but this was the first time that responsibility had been pinpointed on one individual at local director level. One advantage of such an appointment was that it avoided the early involvement of the chief executive when a major dispute occurred and at least postponed the development of a situation which could be presented as a personal confrontation between the chief executive and the strike leader. In the 1968 dispute it had been all too easy to report in terms of Redshaw v Montgomery, the district secretary of the AEF, with admitted Communist sympathies.

When the Board came to discuss the loss of £768,000 by Hydraulics Division, closure of the Division clearly had to rank among the options. Had the loss been an aberration of some kind it would have been another matter, but the viability of the business had been in doubt over a long period. In 1967 the loss had been £332,000 so that in two years results from the Engineering Group had been pulled down by £1m. 'Well, what can you expect if you keep moving it from works to works?' asked one director in effect. Certainly, the more recent results reflected in part at least the disturbance involved in the move from Weymouth to South Marston in 1966. Before that there had been two other moves—from Crayford to Elswick and from Elswick to Weymouth. In each instance, as part of a larger reorganisation, cogent reasons existed for the move, but any benefits gained in other ways had to be set against the disruption caused to a business which at least in theory had considerable potential. Hydraulic equipment was indispensable to modern engineering: Vickers knew all about hydraulics and had been in the business for decades: why then could they not make it pay? At the root of the matter lay intense competition. The market might be vast and more likely to grow than diminish, but there were too many flies round the honey pot and few garnered more than meagre sips, especially in years when the honey flowed less freely. With competition as keen as this, not only had products to be first rate, but they had to be available at a price which barely covered the costs in developing them. The Vickers business, it was variously suggested, did not offer a wide enough range of products nor enough new products: it was too small to compete with the international giants and it should be either greatly expanded or discarded. Discarding was all very well, pointed out J. H. Robbie, the Engineers' chairman, but withdrawal would be very costly in view of obligations undertaken on spares and repairs and of contractual arrangements with the American company, Racine. The Board decided rather unhappily that the business must continue, but called for further vigorous efforts to cut costs and improve efficiency. To the extent that the loss in 1969 was £262,000 these efforts produced an improvement and the improvement continued into 1970 when a small profit was achieved. The fact remained that Hydraulics was still a business earning inadequate returns, and had the cost of withdrawal not been so heavy a decision to close, sell or merge would almost certainly have been taken. Influencing the decision was the future of the South Marston Works as a whole since in terms of capital employed hydraulics was the biggest activity located there.

Another engineering activity which did not prosper as had been hoped by transfer to South Marston was All Wheel Drive. This had become the central activity at the works of Vickers-Armstrongs (Onions) Limited at Bilston. Principally it involved the conversion of trucks to 4-wheel drive and the building of special chassis for crane carriers. In 1968, as part of the general programme of rationalisation in the Engineering Group, the Bilston Works was sold and these activities moved to South Marston in the hope that the environment there would prove more congenial. It was a hope quickly found to be vain. As Vickers All Wheel Drive at South Marston it fared little better than it had at Bilston and following losses in 1968 and 1969 it was sold to the George Cohen 600 Group in 1971.

In these and other ways the Engineers were going through a rough patch, and they might have felt that 1968 was hardly the right moment to move Barrow Engineering Works, their biggest unit and profit maker, into the Shipbuilding Group, whatever good sense this made in terms of general policy. In 1967, with Barrow, they showed a profit of £1.274m. Without Barrow it would have been a loss. Not surprisingly, then, they slipped into the red in 1968 to the extent of £132,000 and could manage a profit of only £477,000 in 1969. It seemed that they were fated to lose their strongest runners. In 1965, with the Crabtree acquisition, printing machinery had been given a separate status. In 1966, with the Roneo acquisition, the Dartford activity in office furniture had been transferred to the new office equipment group.

They had suffered a blow of a different kind in 1967 when the Minister of Defence (Equipment) announced that the whole of a further order for Chieftain tanks was to go to the Royal Ordnance Factory at Leeds. Moreover, the Ministry had indicated that with completion of this order no requirements could be foreseen for tanks for the British Army for many years to come. So the future of the Armament and Commercial Engineering Division at Elswick was once again in doubt with all this implied for Elswick Works as a whole. The situation was summarised in an announcement by the Company in April 1968. 'Work on the existing Chieftain order will be completed during the first half of 1969,' it said, 'and the future of Armament Division will then depend on the degree of success achieved by present efforts to obtain orders from overseas for the Vickers Battle Tank, other tracked vehicles and component supplies.' Like Pearl White, the Division was once again rescued in the nick of time. After a long wooing, the Government of Kuwait placed an order for Vickers tanks, to a value of some £7m, and so continuity of work was secured by this and other export orders.

After the floods at Dartford in 1968, Roneo Vickers suffered again in 1969 from an 'act of God'. In 1968 it had been water: now it was fire. Not one but two fires occurred at Romford Works, the main manufacturing centre. The first and more severe occurred in March, destroying the section of the Works producing supplies and systems, an area of 95,000 square feet. As at Dartford, management and work force accepted the disaster as a test of mettle and performed prodigies in restoring production. In the meantime, as the annual report recorded, 'stencil supplies were obtained from other office equipment companies, both at home and abroad, who did their utmost to help us out of our difficulties'. In a highly-competitive industry, this evidence of camaraderie is worthy of remark. The second fire took place in December, destroying 50,000 square feet of warehouse space and a considerable quantity of materials, but again the dislocation was minimised by vigorous improvisation while rebuilding took place. Financially no great loss was suffered as insurance existed to cover both damage and consequential loss.

For the Office Equipment Group 1969 was, in fact, its most successful year since its formation in 1966, with profit topping £2m from sales totalling £28m. This contribution to Vickers pre-tax profits of £8.41m for the year was second only to that of the Shipbuilders. For the third successive year pre-tax profit showed a substantial increase and it seemed that at last Vickers was well and truly on its way to better things. A journalist had written in 1968: 'Problems plague Vickers like fleas do dogs. No sooner does the company get one off its back than another appears.' In 1969 they were back again, but before then much else had happened.

7

Pruning Season 1965–70

If proneness to accident was a feature of the second half of the sixties, so was pursuit of the twin policies of eliminating activities that were either unprofitable or no longer 'fitted' and of releasing assets surplus to need. In line with this thinking opportunities were also sought to dispose of minority holdings in other enterprises.

Scope for rationalisation was particularly apparent in the Engineering Group. Three of its major works – Elswick, Scotswood and Crayford – had been built on a very large scale in the far distant past to serve purposes no longer existing. Elswick, for example, had at one time included warship building in its activities. Crayford had been an immense munitions factory during the First World War. Scotswood had originally been focused on locomotive building. In the context of present activities all made lavish use of space. If this space could be used more economically overheads could be reduced and considerable areas made available for disposal. South Marston also had a space problem, but here it was the need to fill space usefully. Built as an aircraft factory at the beginning of the Second World War it sprawled over an area of 40 acres, without counting the airfield which covered another 544 acres. Space to spare at South Marston, in an industrial development area with excellent communications, represented a strong inducement to rationalise by transferring activities from smaller works which could then be sold. Hence the decisions to take hydraulics from Weymouth and All Wheel Drive from Bilston. If the moves did nothing to improve performance of the two businesses it was at least neat and tidy and brought in some £580,000 from sale of the premises.

At Newcastle the possibility that the Armament Division would have to go out of business had in 1967 pointed to closure of Elswick and concentration at Scotswood. An upheaval on this scale was averted, but plans went ahead to concentrate activities into smaller areas at both works. Property was thus released which in due course became vested in Vickers Properties Limited, a company set up in 1972.

It included the entire river front area at Elswick, with a floor space of nearly one million square feet.

Similarly at Crayford the process of concentration led to a major part of the site becoming surplus to need in 1968. In the event, it proved possible to sell most of the site, with an arrangement for leasing back for five years the portion required by the Engineers. A breathing space was thus provided while the longer-term future of Crayford activities was worked out. The sale realised £2m which, added to the amounts received from other sales, in particular those of Weymouth Works and Bilston Works, brought the total received from disposals during this period to some £3m. Except in cash flow terms, however, receipts of this kind did not necessarily represent a gain to the Company. A sale might be made at a price exceeding historic book value, but its relationship to current book value could be a different matter, and in any case there had also to be taken into the assessment the costs of reorganisation. The practice had been established in 1966 of charging reorganisation expenses to reserve, and similarly of crediting to reserve sums from the sale of fixed assets in excess of book value. At the end of the day there was probably not a great deal in it.

For the Shipbuilding Group Palmers Hebburn was the problem child. The site on the south bank of the Tyne had been used for building wooden frigates for the Royal Navy from the beginning of the 19th century. In 1851 two brothers, Charles (later Sir Charles) and George Palmer, took over the yard and began the development of iron shipbuilding, with HMS *Terror* in 1854 as the ship which introduced rolled steel armour plates. The firm remained in being until the slump after the First World War. By 1932 things had become so bad that they had to sell out to the National Shipbuilders Security Limited who decided to keep them going as shiprepairers only. Two years later Vickers bought the business and spent £100,000 on a programme of reconstruction. By the outbreak of the Second World War the yard had excellent facilities, including the largest privately-owned dry dock on the east coast. Throughout the war it worked at full stretch on naval conversions, repairs and refits. The company became a unit of Vickers-Armstrongs (Shipbuilders) in 1958, and in the following year work began on a dry dock capable of handling the new generation of super tankers. The total cost of the dock and other modernisation work was £4.5m and this was incurred without benefit of the development grant which would have been available a few years later. The official opening took place in September 1962, and by then the dock was already more in the nature of a white elephant than a compelling competitive asset. It had been designed to take 100,000-ton tankers. In the aftermath of Suez tankers had escalated in size and the super tanker was now more likely to be 250,000 tons than 100,000. Moreover, repair facilities had been created at places more conveniently situated on the main trade routes. Palmers' problems were compounded by the fact that amortisation of the cost of the new dock added to their overheads and so reduced their competitiveness. In these circumstances annual results were consistently poor and, with no prospect of the circumstances changing, Vickers eventually felt that the nettle must be grasped. With 1,100 jobs at stake there had been a natural reluctance to take drastic action. Efforts to find a buyer, including an approach to Swan Hunter, had been unsuccessful. Closure was the last resort, but for those who lost their jobs redundancy payments would provide some alleviation.

The Company's intention to close the yard by the end of June 1970 was announced in April. Advance warning had been given to the Ministry of Technology. The Minister (Mr Benn) responded by asking for the closure to be postponed for some months in order that the Government could have time to study a report pending from the Shipbuilding and Shiprepairing Council on the state of the shiprepairing industry as a whole. Since a delay would impose additional losses on Vickers, he accepted the argument that a contribution should be made towards these losses. While doubting if the situation could be significantly changed, Vickers accepted the Minister's proposal in principle, and it was agreed in discussion that the closure should not take place until the end of September, with the Government providing up to £200,000 towards losses during the extended period. Unhappily, when the report arrived it did not change the basic situation, and with the Government indicating that they had no further proposals to offer arrangements went ahead for closure of the yard by the end of September. By then the number employed had fallen from 1,100 to 750 through voluntary movement. Though shiprepair work ended this did not mean closure of the galvanising and gritblasting business which had been established on the site and was administered by the Shipbuilding Group. On the contrary, further investment was made in the business by installing the largest ceramic galvanising bath in the United Kingdom, holding over 1,000 tons of molten zinc and capable of galvanising steel units up to 70 feet long and six feet wide.

With the possibility in mind that at some later date Palmers Hebburn might yet be sold to a shiprepair or shipbuilding organisation the facilities were kept in being. These hopes materialised in mid-1972 with purchase of the yard by the Swan Hunter Group. The consideration was £1.335m in cash, together with 129,999 ordinary shares held by Swan Hunter in Brown Brothers & Company Limited. It was a reasonably satisfactory ending to a disagreeable episode in that Vickers lost less than they would have done, Swan Hunter added usefully to their facilities and the Tyne employment situation was improved. Also for Vickers was the advantage of now holding all the ordinary shares in Brown Brothers, 'marine and general engineers operating from Rosebank Iron Works, Edinburgh', and employing 700 people. Browns, a long-established firm, had a high reputation as suppliers to the shipbuilding industry, in particular of steering gear and ship stabilisers. They now became part of the Shipbuilding Group.

Another business closed in 1970 was Ioco Limited at Glasgow. It had first come into Vickers orbit as the Imperial Oilskin Company when Vickers began building airships just before the First World War. By 1918 Ioco had produced over 2,500,000 yards of fabric for this purpose. Because of these close links Vickers took a shareholding in the company in 1911 and then acquired it as a subsidiary in 1919. In the 1920s the Aircraft Guarantee Company, a Vickers subsidiary, built the R100 with Barnes Wallis as the designer and Nevil Shute as his chief calculator. The R100 flew successfully to Montreal and back in 1930, but shortly afterwards the R101 (built in the Royal Airship Works at Cardington) crashed in France on a flight to India, with the Secretary of State for Air on board, and this disaster brought a sudden end to British interest in airships. Ioco had other outlets for its products, however, including the electrical industry which provided a growing market for Ioco insulating materials. In the 1960s it had a range of products from insulation materials and laminates for mechanical engineering to rubber coverings. The business was always struggling, but the characteristic reluctance in Vickers to close a business led instead to further investment in modern plant. This did little to improve the position or to encourage possible buyers, and with accumulated losses of £650,000 in seven years' trading the decision to close was taken with effect from the end of 1970. The closure involved some 300 redundancies. The premises were sold in 1972.

In a different type of disposal operation the operating subsidiaries of the Olding Group were sold in 1969. Though profitable, the business was considered no longer to 'fit' into the pattern of Vickers activities. Its history as a Vickers company went back to 1953 when it served as the distributing agency for Vickers' tractors and a controlling

interest was acquired. Since Oldings had franchises from other manufacturers of earth-moving and allied equipment, they survived collapse
of the tractor project and operated successfully on sales of products
from leading companies such as Clark, Barber-Greene and Massey-
Ferguson. In 1967 and 1968 trading profit was in the order of £200,000.
With the Wiles Group Limited appearing on the scene in 1969 as
an interested buyer, Vickers decided to sell and to pave the way
bought for £805,000 the minority holding in the parent company.
The principal operating subsidiaries were then sold to Wiles for
£1.570m against a book value of £2.150m. Assets remaining in
the Olding Group comprised £2.1m in cash and a 49 per cent investment in Barber-Greene England Limited. The original cost of the
Barber-Greene investment had been £255,000, but its current value,
said the Directors' Report, was considered to be substantially greater.
How much greater became apparent in 1976 when the holding was
sold for £2.250m to Barber-Greene Company, of Aurora, USA,
Vickers' partners in the business.

The pruning extended to investments on the grounds that money
should not be locked into companies whose policies Vickers could
not control or at least strongly influence. Two minority holdings
had arisen from efforts to withdraw from unprofitable and unpromising
situations – those in hovercraft and at Naval Yard. The Tyne merger
in January 1968 left Vickers with a holding of 16.36 per cent in Swan
Hunter Shipbuilders Limited. In December 1969 this holding was
sold to Swan Hunter at its book value of £1.168m. In hovercraft
the merger with Westlands in 1966 gave Vickers 25 per cent of the
share capital of British Hovercraft Corporation. In 1969 and 1970
discussions took place between the three partners (Westlands, Vickers
and NRDC) about the injection of additional capital. The proposition
had no attraction to Vickers who made it clear that they would
prefer to disinvest. This did not prove easy to negotiate, but in October
1970 it was agreed that Westland would buy the Vickers holding.

On an entirely different scale of magnitude was the Company's 12.64
per cent stake in International Computers (Holdings) Limited. Vickers
had been a party to the 1968 merger on the understanding that it
would be free to dispose of its shareholding in the enlarged company
if and when a suitable opportunity occurred. The suitable moment
proved to be immediately after the annual general meeting of ICH
in January 1970. The holding was offered on the market and brought
in £10.481m against a net book amount of £9.336m.

In completing its severance from hovercraft, and from tabulators
and computers, Vickers could look back on a significant contribution

to the development of both activities in the post-war period. Continuing in the computer business would have meant plunging into electronics in competition with established giants, and this would have made little sense for a company whose expertise lay essentially in mechanical engineering. As for hovercraft, Vickers had been successful pioneers: they had designed and built vehicles which demonstrated the practicability of the concept, but the market had developed too slowly and on too limited a scale for successful pioneering to be rewarded by successful commercial exploitation.

Stretching even further back was the association with Metropolitan-Cammell (Holdings) Limited, jointly owned by Vickers and Cammell Laird. Its activities lay in railway rolling stock and motor buses, in direct line of succession from the stage coaches with which the business had been started in pre-railway days by Joseph Wright, a London coachbuilder who also operated a stage coach service between London and Birmingham. With the coming of railways, and operating on the principle of joining them if you can't beat them, he began designing and manufacturing rolling stock in the early 1840s, and for this purpose built a factory on a green field site in the Saltley district of Birmingham. The business prospered and expanded and in 1902 was incorporated as the Metropolitan Carriage, Wagon and Finance Company. During the First World War its chairman was Dudley Docker, a man of immense drive, energy and ambition, who believed that the modern world would be run largely by electricity so that one had better get involved in electrical as well as mechanical engineering. In 1917 Metropolitan Carriage accordingly acquired a controlling interest in British Westinghouse Electrical and Manufacturing Company. Douglas Vickers knew Docker and shared his view about the importance of electrical engineering, to the extent indeed that Vickers bought from Metropolitan Carriage 45 per cent of its holding in Westinghouse to add to several other electrical interests, including a traction motor factory and a business engaged in electrical heating and cooking appliances. Two years later, in 1919, Vickers bought Metropolitan Carriage itself and so was now strongly involved in both railway rolling stock and electrical plant, with the British Westinghouse subsidiary re-named Metropolitan-Vickers Electrical Company. The cost to Vickers was £13m, a very considerable figure at that time and one that caused critical comment. Under the strains and stresses of the great depression of the 1920s it did not take long to decide that a mistake had been made. Immediately following the upheaval which brought Vickers and Armstrongs together in 1927 Vickers sold a controlling interest in Metropolitan-Vickers to the International

General Electric Company of the United States (leading to the formation of Associated Electrical Industries Limited). The rolling stock
business of Metropolitan Carriage was retained, but in 1928 Vickers
and Cammell Laird agreed to bring together their rolling stock
interests and to this end formed a new company, jointly owned, under
the title of the Metropolitan-Cammell Carriage, Wagon & Finance
Company. In much the same way a year later the two companies
merged their steel interests to form English Steel Corporation, though
in this instance on a 75–25 basis.

In its new guise the rolling stock company made a very poor start.
It depended heavily on exports, and more and more countries were
beginning to manufacture their own rolling stock. Additionally, this
was a time of economic crisis. So serious did the situation become
that in 1934 the company was put into liquidation and a new company
formed. It had almost the same title, Metropolitan-Cammell Carriage
& Wagon Company Limited, and as before Vickers and Cammell
Laird held the shares jointly. Later in the thirties the trading situation
improved and during the Second World War the factories at Birmingham worked at full stretch. Ready markets were also available during
the period of post-war reconstruction. With its products held in
high repute, Metropolitan-Cammell again exported on a large scale,
especially to Commonwealth countries. Buses and bus bodies were
now also part of its stock-in-trade, and these found a particularly
big market in the Caribbean area. Exports were well supported by
home orders, including notably 532 tube cars for the Piccadilly Line,
followed by 450 for the Victoria Line. In 1958 profits reached nearly
£1.5m. In the early sixties the picture changed dramatically and in
1961 the company had a loss of over £250,000. Markets had shrunk
and become severely competitive so that profits fell sharply even
when the order book looked strong. British Railways now built
more of their own rolling stock. Metropolitan-Cammell responded
by reorganisation, by disposing of part of its manufacturing facilities
and by a programme of diversification, including containers. In
these circumstances, divided rule, represented by a 50–50 partnership,
was not always comfortable and at the end of 1965 the Cammell
Laird chairman, P. B. Hunter, suggested that either one of the parents
should take full control or that one should take over the rolling stock
division and the other the bus division. In further discussion it emerged
that Vickers was willing to sell and Cammell Laird to buy, and the
two partners eventually agreed on a price of £1.750m. Vickers
interest in Metropolitan-Cammell thus ended on 1st January 1969,
after fifty years.

Other pruning in the second half of the sixties included closure of the steel foundry at Barrow and the Constructional Steel Division at Palmers Hebburn. The industrial pump business at Barrow was sold, and at Elswick activities discontinued in marine equipment, forge and die production. Nor did Vickers House escape. Reductions in Head Office staff led to release of seven more floors at Millbank Tower for letting to outside tenants, and Vickers occupation was now restricted to six floors.

Writing of these developments, Sir Leslie said in the 1969 annual report: 'Some of the proceeds have been used to pay off debts already incurred in making acquisitions: but clearly our ability to invest further in growth activity has been enhanced.' A list of acquisitions and disposals during the five years to March 1970 was published in the report. Proceeds from disposals, it was recorded, totalled £32m and the cost of acquisitions over £31m. Of the receipts from disposals all but £6m came from steel compensation and from sale of the investment in International Computers.

The main acquisitions had been those in 1965 in printing machinery, office equipment and chemical engineering, followed by the purchase of Algraphy Limited in 1969. Including Algraphy, the cost of acquisitions in printing machinery and supplies totalled £15m. In office equipment the total was £12m and in chemical engineering £2.2m.

Smaller acquisitions were those of Compact Orbital Gear (Rhayader) and Slingsby Sailplanes Limited (Kirkbymoorside, Yorkshire) by the Shipbuilding Group, Michell Bearings Limited (Newcastle), in which Vickers already had a minority holding, and Kirby Engineers Limited (Walsall) by the Engineering Group and Hadwa Limited (Perth, Australia) by Vickers Australia.

Of these five, Michell Bearings, after a period of travail, was to emerge in the seventies as one of the most successful businesses in the Engineering Group. Also to be ranked as a 'good buy' in the light of subsequent developments was the Hadwa steel foundry in Western Australia: not least it was well sited to supply spares to the mining industry.

The acquisition of Slingsby Sailplanes, a leading glider manufacturer, caused some raised eyebrows among those who could see no apparent connection between shipbuilding and sailplanes except for Leonard Redshaw's enthusiasm as a glider pilot. In fact, his interest had been roused by Slingsby's link with a German firm, Glasflugel, possessing a high degree of knowledge in glass fibre technology. Through this link Redshaw saw the possibilities of developing applications of glass fibre technology within Vickers.

including use in submarine construction. Eight years later the world's first glass reinforced plastic submersible went into service in the North Sea with Vickers Oceanics, built on the Yorkshire moors by Vickers-Slingsby.

8

City Intervention 1969–70

Growing intervention by Government (creeping intervention, some called it) had become a part of life in industry generally, and in Vickers in particular, during the 1960s, but at the end of 1969 Vickers found itself confronted also by an intervention from the City. This was the first overt intervention in the affairs of a major industrial company by a group of institutional shareholders. Vickers was the guinea pig and the outcome of the experiment was bound to effect the course of future relations between the City and industry.

In ignorance of what was brewing, Sir Leslie Rowan may well have felt in the summer of 1969 that now at last the worst was over. His two years in the chair had been in rough seas, but perhaps now a course had been set into smoother waters under sunnier skies. New activities showed encouraging signs of doing well and 1969 seemed unlikely to produce another of the sudden 'accidents' which had so bedevilled the Company in recent years. Profit before tax had risen from the very low figure of £3.94m in 1966 to £4.68m in 1967 and £6.98m in 1968, and projections for 1969 pointed to a 50 per cent increase for the third successive year. True, the 10 per cent dividend had not been covered for three years, but cover was now expected. As the old guard in senior management approached retirement younger men, capable, sharp-minded and ambitious, were beginning to emerge. From outside J. S. Booton had been recruited as Comptroller and was working closely with I. P. Coats, Controller of Group Planning, also a younger man in the upper reaches of the hierarchy. Of the 120 senior executives throughout the Group, over half were under 50 and 15 per cent under 40. At Board level, the succession to the chairmanship seemed to have been secured by the announcement in May that Mr Niall Macdiarmid had agreed to become an executive director and Deputy Chairman from the beginning of 1970. By then his resignation would have taken effect from the British Steel Corporation where he was managing director of the Northern and Tubes Group: before that he had been chairman of Stewarts & Lloyds, one of the major businesses caught up in steel nationalisation.

It had certainly been a difficult two years and anxieties still abounded, but the sense of being on the edge of crisis was less obtrusive. Speculation about a possible take-over had died down, for the time being at least. In 1968 the *Economist* had written of Vickers as 'the most gossiped-about take-over prospect in Britain', adding that efforts to move out of 'production-oriented armaments' had been 'watched with a mixture of cynicism, sympathy and greed in investment circles'. A company was particularly vulnerable, the *Economist* pointed out, when it had put itself into a recovery position. On the 29th floor at Vickers House the possibility of a take-over was recognised, but on the whole discounted. The very large engineering companies, such as GKN and Tube Investments, had indicated no interest and the danger lay more with asset strippers – take-over followed by 'profitable dismemberment', to quote the *Daily Telegraph*. At one stage the grapevine said that a close examination was being made of Vickers by James Slater, then a highly active predator, but no overt moves were made, and no significant movement of shares could be detected. Nevertheless, a state of alert was maintained and Sir Leslie knew that any suggestion of interruption in Vickers recovery, or any recurrence of 'accidents', could quickly provoke a crisis.

It was for this reason that the half-year results for 1969, published in September, caused him some alarm. They showed pre-tax profit of £3.6m compared with £4.9m for the first half of 1968, and the City might well have concluded that there was little prospect of achieving in the full year the 'advance in both profits and return on capital employed' which Sir Leslie had forecast at the annual general meeting in June. They might well conclude also that for the fourth successive year full cover would not be provided for a 10 per cent dividend. Internally, there was confidence that the better figures would be reached, and the half-year statement included reassurances. 'The flow of completions in 1969 in certain important activities is such that profits will largely accrue during the second half of the year,' said the statement.

The fact remained that a lot would have to be done in the second half of the year, and especially in the last quarter, to meet targets. The importance of giving the highest priority to this effort was therefore stressed to the chief executives of the operating groups. In discussion at a meeting of the Managing Director's Committee attention was drawn to the competing claim on senior management time involved in preparation of the 1970 management plan. The established timetable for management plans required their submission in October and November for challenge and discussion so that the final plan for the Group as a whole could be placed before the Board in January. Obviously work on the plans occupied a good deal of senior management time and to require them at the same time to pull out all the stops to maximise profit in 1969 meant that they would be under very heavy stress during the last few months of the year. Feeling that he should do something to alleviate the situation, Sir Leslie looked again at the management plan schedule and decided that its stages could be telescoped in such a way that the pressure could be taken off management during the crucial final quarter without impeding the purposes of the plan. These purposes he related to the need to have full and accurate information at the times when the Company was required to make public statements – in the annual report and in the half-year statement. On this thinking he concluded that it would be sufficient for the plans to be submitted by the 1st January, instead of in October and November, for the corporate plan to go before the Board in March, and for a single forecast of results to be made once a year in September, instead of three (April, June and September). These ideas he put into a minute as a basis for discussion, emphasising his view that 'the Management Plan system and discipline are essential and we must do nothing to undermine their authority'.

The minute brought an unexpectedly strong reaction from the two executives at Head Office principally responsible for planning procedures and for making use of the data contained in the plans – I. P. Coats on the planning side and J. S. Booton on the financial side. Jointly they produced a long memorandum with extensive supporting documents and tabulations. In essence the memorandum first set out the arguments against amending the schedule in the way proposed by the Chairman and secondly made recommendations for reconstructing the Board and for separating the roles of chairman and chief executive (in existing practice the chairman was also chief executive). The schedule proposed by Sir Leslie they considered to demonstrate a wrong conception of priorities and to cast aside all the arguments and thinking that had led to formulation of the original schedule in 1964, with advice from McKinsey, and to subsequent refinements. They felt that what he now proposed would bring into doubt and disrepute not only their own work and purpose, but also group management as a whole. They spoke also of 'contradictions' in Sir Leslie's minute and referred to his 'professed' belief in the value of the management plan system and discipline. On the question of composition of the Board they argued for a complete separation of the functions of determining

11

policy and executing it. On this proposition they thought that the Board should comprise only the chairman, the chief executive and the non-executive directors. Below the Board there should be a management company to advise and assist the chief executive. Its chairman would be the chief executive and its directors representative of the operating groups and the service functions. Entailed in this proposition was separation of the roles of group chairman and group chief executive.

Sir Leslie was considerably upset by this memorandum, perhaps less by the arguments and concepts it advanced, which could be debated, as by its tone and the attitude of hostility he felt it revealed. As a check on his own reactions he asked for the views of other members of the Board individually. A clear consensus emerged that, whatever its merits, the memorandum recorded lack of confidence in the Chairman, in the Board and in top management and that this was a situation which made it virtually impossible for the authors to continue to hold high office in the Company. The upshot was an internal announcement on the 7th November that 'by mutual agreement Mr J. S. Booton and Mr I. P. Coats will relinquish their appointments in the Company on 31st December 1969'. It could not be expected, however, that the episode would remain one of internal interest only, and it quickly became apparent that the matter was being talked about in the City. In an article in *The Times* Robert Jones commented: 'The departure of the two men who in the City were regarded as the vanguard of the forces of change will not help confidence.'

This clash may or may not have precipitated the intervention, but certainly 'the trouble in Vickers' was mentioned when the opening shot was fired a few days later. The man with the gun was Sir Kenneth Keith, chief executive, Hill Samuel & Company Limited. In conversation with Colonel Maxwell at a dinner party on the 10th November, and emphasising that he was speaking as an investment manager and not as a merchant banker, Sir Kenneth said that he and others in the City were worried about Vickers management, wondered whether the Board was sufficiently 'with it' and questioned whether Niall Macdiarmid, though held in high regard, was the right man to become chief executive in these particular circumstances. When these opinions were relayed to Sir Leslie he took the view that if they were seriously intended they should be expressed to him directly as Chairman.

A formal approach followed on the 17th December when Sir Kenneth came to see Sir Leslie. He stressed that he came in an entirely helpful spirit, but he and three other major shareholders – the Prudential Assurance Company, Cables Investment Trust and Britannic Assurance

Company—were worried by Vickers' record and prospects and believed that changes should be made in the Board and in management at the top. They thought that new, tough outside directors should be brought in quickly, having the confidence of the City and ready to wield the axe. Their role would be somewhere between that of an executive and of an outside director and they would be available on a part-time basis to give advice and assistance.

Sir Kenneth said that the question of Vickers had first been raised with him by Mr Ronald Grierson, and that Mr Arnold Weinstock and Sir John Mallabar had also taken part in subsequent talks. Mr Grierson, whose background also lay in merchant banking, had been deputy chairman and managing director of the Industrial Reorganisation Corporation from 1966–67: Mr Weinstock was managing director of GEC: and Sir John Mallabar senior partner of J. P. Mallabar & Company, chartered accountants, and at this time chairman of Harland & Wolff. He did not, however, suggest that they were involved in the approach he was now making. In fact, Mr Grierson had come to see Sir Leslie a few months earlier to raise the possibility, apparently as a *ballon d'essai*, that he and a number of other individuals might acquire a substantial shareholding in Vickers, with the implication that they would then expect to have a correspondingly strong voice in conduct of the Company's affairs. The matter had not been pursued on either side.

At the initial meeting with Sir Kenneth no more was sought than to lay cards on the table and a fuller discussion took place at a lunch a fortnight later with Sir Kenneth as host. With him were Mr Angus Murray, of the Prudential, and Mr John Pears, of Cable & Wireless, and Sir Leslie was accompanied by Mr Yapp, and by two outside directors, Sir Peter Runge and Mr Faulkner. The Britannic was not represented at this or subsequent meetings. The position was re-stated much as before, with reference to the pressure being put on the institutions to take a more active interest in company affairs. For his part Sir Leslie expressed surprise at the timing of the intervention since he believed that much 'surgery' had taken place or was in prospect, profitability was moving upwards and prospects were good. In these circumstances, he suggested, it would be in the interests of everybody to avoid rocking the boat. Both sides agreed that publicity should be avoided. Sir Leslie also pointed out that five executive directors, including himself, would reach retiring age during the next three years and that this had been seen as the opportunity to bring in a new executive team. At this point Mr Yapp interjected to say that if it would be helpful to the future of Vickers, by facilitating management

change, he would be perfectly ready to retire ahead of normal retirement age. What Vickers needed, it was suggested, was more 'sons of
bitches' in the organisation. Dignity and wisdom were not enough
and some 'nimble East End qualities' would be more appropriate to
the situation.

Assessing the situation in the light of these talks, the Board concluded
that the approach by the institutions must be taken seriously. The
share register showed the Prudential and Prudential nominees as
holders of some two million pounds of ordinary stock which made
them the biggest single holder. The precise holdings of the others
involved was not necessarily revealed by the share register because of
nominee holdings, but together the Prudential, Britannic, Cables
Investment Trust and Hill Samuel probably held 10 per cent of the
equity. It was likely, however, that they could control at least 20 per
cent and would be able at the annual general meeting to bring about
removal from the Board of those directors of whom they disapproved.
Against this background the Board agreed that action should be put
in hand to seek additional outside directors and to find a new Managing
Director, given Mr Yapp's declared willingness to relinquish the
appointment.

This was the position when during the second half of January Sir
Kenneth said that he wished to bring the Industrial Reorganisation
Corporation into the talks, in the persons of Sir Joseph Lockwood, the
chairman, and Mr Charles Villiers, the managing director. To Sir
Leslie this seemed a very odd development since the IRC was a parastatal organisation, with no stake in Vickers. The reason for the move,
Sir Kenneth explained, was that while the institutional shareholders
had power they did not have knowledge of industrial management
or of the inner workings of Vickers: the IRC would bring this knowledge to the talks. The extent of IRC knowledge of Vickers was
regarded in the Company as being extremely limited: a willingness
to talk to the IRC about matters in which there might be co-operation
had been reflected at an earlier stage in three short visits to Vickers
establishments by representatives of the IRC, but this, it was felt, did
not constitute 'knowledge' of the kind that the institutional group was
seeking. Nevertheless, the group was entitled to take advice from where
it saw fit and a meeting took place with Sir Joseph Lockwood and Mr
Villiers. The meeting, in Sir Leslie's view, was highly disagreeable. It
was followed by a further meeting with the institutional group, plus
Sir Joseph and Mr Villiers, when Vickers were pressed to separate the
offices of chairman and chief executive, as a number of other companies
had done or were doing.

The situation took a disconcerting twist a week later, on the 12th February, when Patrick Sergeant, City Editor of the *Daily Mail*, prominently published news of the talks under the heading 'The heat's on Vickers now'. The article said forthrightly: 'One of the largest merchant banks is leading a group to ginger up Vickers. They want to make changes in the management and in the policy.' For Patrick Sergeant to get a scoop was nothing new and it was no use bewailing the fact that he had done so again, but Vickers could only suspect that in order to put pressure on them there had been a 'leak' from the institutional group, directly or indirectly, deliberately or not, though this was denied. It would make subsequent talks less easy to conduct and would create unease and uncertainty throughout Vickers, but the effect on share prices was unlikely to be harmful.

In order that the Board should know at its February meeting exactly what action the institutional group expected Vickers to take, Sir Leslie asked Sir Kenneth to set this out formally. This he did by letter. It was essential, he said, that a managing director should be appointed with 'wide executive powers' and 'responsible under the authority of the Board for the initiation and formulation of the commercial and financial policy of the Company': finding the right man was the responsibility of the Board and would not be easy though names could be suggested from various quarters: if the Board did not agree consideration would have to be given to discussion with other large institutional and private shareholders about the means of effecting changes in the Board in order to ensure that the appointment was made.

To an important extent this cleared the air, and the consensus view of the Board showed a willingness to accept that there was a strong case to be made on its merits for separation of the offices of chairman and chief executive. The task was therefore to find the right man as chief executive, highly difficult, as Sir Kenneth had said, since one had to expect that people of this calibre were few and far between, were much in demand and were rarely available at the moment they were sought. For the rest, it seemed that the proposal for the appointment of 'semi-executive' directors had been dropped, though Sir Leslie continued to look for suitable people to be invited to become ordinary non-executive directors. It also seemed that the IRC would no longer be involved in direct talks between the two sides, and to the extent that their presence had been regarded by Vickers as inappropriate future discussion might be expected to proceed more smoothly. There had been some second thoughts, it seemed, in the institutional camp, and the doves had overcome the hawks.

Within Vickers the three people most directly and personally concerned were Sir Leslie as Chairman, Mr Macdiarmid as Deputy Chairman and Mr Yapp as Managing Director. At the outset Mr Yapp had declared his willingness to retire before 65 if this would help Vickers to resolve the situation now facing them. The Board nevertheless felt that by reason both of his personal qualities and of his intimate knowledge of the Company's affairs, not least on the aviation side, his services should be retained if possible. Accordingly it was proposed to invite him to become a non-executive director in the event of his resignation as Managing Director, and that he should continue to be a non-executive director of BAC. For Sir Leslie, loss of the function of chief executive would obviously represent a major readjustment, though as non-executive Chairman he would continue to preside over the fortunes of the Company. For Mr Macdiarmid the position was perhaps rather more complicated. He had duly joined the Company at the beginning of 1970 as executive Deputy Chairman, with the expectation that in due course he should succeed Sir Leslie as Chairman and chief executive. Now he could still expect to succeed to the chairmanship but in a non-executive capacity only. Why not, it was suggested, appoint him instead to the new office of managing director and chief executive? Against the known views of the institutional group, however, this could not be seen as a realistic possibility.

In fact, it was Mr Macdiarmid who proposed for the list of candidates the name of the man who was eventually invited to accept the appointment. This was Mr P. A. Matthews who had been a senior executive of Stewarts & Lloyds at the time of Mr Macdiarmid's chairmanship and was now a Member of the British Steel Corporation, and Managing Director, Operations and Supplies. At 47 he was in precisely the right age bracket and his career showed a range of experience admirably suited to the task of taking over the executive leadership of a major industrial company. He had both technical and sales experience and his senior appointments had included those of director in charge of planning and development engineering with Stewarts & Lloyds and a group technical director of BSC with responsibility for research and development, engineering and planning and management services. He had been born in Canada and after receiving his early education there went to Oundle. On the outbreak of war he was commissioned in the Royal Engineers, with whom he served in North Africa, Sicily, Italy and Greece. His subsequent appointments took him to many parts of the world, including most of the world's oil fields.

To have found a man with qualifications as excellent as these was one thing, but there was then the question of his availability. In fact, he was willing to accept the appointment, and the Steel Corporation was prepared to co-operate in his reasonably early release. Sir Leslie accordingly recommended his appointment and the Board approved the recommendation. In May 1970 Vickers announced that he would take up duties at the beginning of September, though in fact he began visiting Vickers works and other establishments in July. The institutional group, through Sir Kenneth Keith, expressed themselves satisfied with the action taken to appoint a managing director with the powers they had prescribed: Mr Matthews was not known to them personally, but they wished both him and Vickers good luck: and the matter could now rest until the new man and his team had had time to prove themselves.

Thus at the end of the day Vickers had that 'little bit of bloomin' luck' which Albert P. Doolittle considered so necessary to the successful conduct of human affairs. They had found the right man in circumstances which were all against their doing so. 'He is by reputation a go-getter, a man of great energy who gets things done without flamboyance, but with remarkable effect,' said the *Daily Mail*.

In terms of public comment, another 'leak' ahead of the announcement had provided the City editors with a chance to make a preliminary assessment of the incident. 'Vickers forced repentance' said the *Daily Mail* headline, and Patrick Sergeant commented that the Vickers Board had traditionally 'seemed more devoted to the national interest than to the shareholders'. He noted also that the share price had advanced, 'partly anticipating the 1969 results and partly on well-informed views that Vickers have agreed to make some important changes'. *The Times* said that this move by a powerful group of institutional investors raised issues of fundamental importance for industry and the City. 'It reflects a strategic change of policy by leading institutions which now accept they must take an active interest in the management of companies in which they invest.'

In the light of this comment Sir Leslie issued a statement to accompany the announcement of Mr Matthews' appointment, and this is quoted in full as an appendix. In essence, he said that a company management must always be ready to listen to the views of the shareholders: when the approach by this influential group of institutional shareholders had been made in December the Board was therefore ready to listen, all the more so since they were already contemplating the need to bring new people into top management: on its merits alone the Board felt that there was a good case for concentrating the powers of chief executive in the office of managing director: the proposition was acceptable both to him and to Mr Macdiarmid:

and in considering the appointment of a managing director with enhanced powers the Board had been aided by the willingness of Mr Yapp to stand down as Managing Director if the Board felt this would facilitate the construction of the new management team. Sir Leslie added: 'No one who has had the privilege of working with Mr Yapp would be surprised by this characteristically unselfish and constructive attitude.'

With more time to reflect on what had happened, commentators showed some disposition to question the merits of interventions such as this.

Ian Davidson wrote in the *Financial Times* that 'it could well be argued that some of the institutions have so much room for improving their own performance that they could hardly claim to be in a position to improve anybody else's'.

Taurus, in the *New Statesman*, commented: 'The disadvantages of such an intervention are first, that it is negative (one can sack the sitting managers without having any clear idea of who ought to replace them), and second, that it is based on a narrow point of view which leaves out half the case: financial institutions, after all, are interested only in a better return on their money and not in social problems which can be conveniently passed over to the State.' It was unlikely, he thought, that there would be many repeats of the Vickers story.

Robert Heller, in the *Observer*, thought there were other places where seniority, security and tradition were marked management characteristics, notably the headquarters of great insurance companies, like the Pru. 'Merchant banks, like all of us, sometimes make mistakes (like doctors, they are good at burying them). And insurance companies also err; for example, they plunged like Gadarene swine into property shares at calamitously high peaks . . . the managerial expertise of the City is not the same as that of the industrial executive suite, and consequently the City's judgment of managers is fallible.'

Making much the same point, *The Director* said: 'This is clearly a sensitive area where the institutions will be wise to move with caution and prudence, bearing in mind the considerable difference between running the day to day affairs of a company and keeping an eye on its longer term profitability.'

For Sir Leslie and Mr Macdiarmid the changed situation in Vickers presented obvious difficulties; nor was it any easier for Mr Matthews. Possibly a successful *modus vivendi* would have been achieved, but Mr Macdiarmid decided to resign at the end of 1970 'for personal and family reasons' and this left open the question of succession to the chairmanship. Almost on cue a new and formidable figure appeared on the scene in January 1971 in the person of Lord Robens, due to complete his 'ten-year stint' as chairman of the National Coal Board in June. Already there was speculation in the press about his future and a number of companies were known to be trying to secure his services: he had, indeed, already agreed to become non-executive chairman of Johnson Matthey, the bullion dealers.

To Eric Faulkner and others it seemed that an immediate effort should be made to bring him into Vickers as chairman-designate. Since the appointment was now non-executive, however, it had first to be established whether Lord Robens would be willing to forego the role of chief executive, or indeed whether he would find it easy to do so even if appointed as non-executive Chairman. After discussion with him doubts were largely resolved. There remained the question of when he should succeed Sir Leslie. By established practice Sir Leslie would normally continue in the chairmanship until retirement in 1973 at the age of 65, but he had told the Board early in 1970 that he would be ready to vacate the chair at any time this would be helpful to the Company as part of the process of reconstruction. Given this attitude the Board concluded (at a meeting from which Sir Leslie withdrew in order that there should be no inhibitions in discussion) that it would be better for Lord Robens to succeed to the chairmanship sooner rather than later, and agreement was reached that Sir Leslie would vacate the chair after the annual general meeting in June 1971. He chose this moment also to retire from the Board, as did Sir Charles Dunphie, who had been a non-executive director since relinquishing the chairmanship in 1967.

Whatever he may have felt about the severance of his links with Vickers, Sir Leslie knew that there was no question of his being 'on the shelf'. The invitations that now came to him included one to become Chairman of the British Council and this he accepted. It was a considerable capture for the Council. The combination of his lively personality and remarkably diverse experience, attainments and contacts was admirably suited to their needs. Unhappily, it was to be only a brief reign. In April 1972 Sir Leslie died suddenly at home at the age of 64.

Part Four
1970–1977

New Broom 1970-71

WHEN HE FORMALLY JOINED VICKERS AT THE BEGINNING OF SEPTEMBER 1970 Peter Matthews had already visited a number of the works and formed first impressions about the quality of the management. He had also had preliminary talks with key figures in Vickers House and formed first impressions about organisation and personalities at the top of the Group. From this base he moved quickly.

The problems requiring immediate attention, he decided, were the potentially dangerous liquidity situation, the absence of a rapid and reliable flow of financial information from the operating units and the depressed state of the Engineering Group.

So far as the liquidity situation was concerned, he identified the most disturbing factor as the largely uncontrolled and unpredictable drain on resources by Vickers-Zimmer. Vickers did not have an adequate understanding of chemical engineering: barriers of nationality and language existed between London and Frankfurt: Zimmers operated mostly on large long-term contracts and to finance them took short-term loans in Germany which Vickers were expected to guarantee: and in the light of experience there could be no absolute confidence that all these contracts would be well drawn and completed without loss. This situation, Peter Matthews concluded, could be resolved only by the earliest possible sale of the business on the best terms obtainable (though even the best terms must represent a substantial write-off).

The task of negotiating the sale fell to a new Director of Finance, Denis Groom, recruited from Rio Tinto Zinc-Corporation Limited in January 1971. Though still only 38, he already had behind him a considerable range of experience in a variety of settings. He was born in London but grew up in Rhodesia, became a Bachelor of Commerce (and a rowing Blue) at the University of Cape Town, was articled in Rhodesia, came to London to practise as a chartered accountant, moved to Newfoundland as Deputy Minister of Finance, and then returned to London to join RTZ as Joint Head of the Group Planning Department. The Groom appointment was supported by that of another young man from the RTZ Group Planning Department, A. B. Butler, who had an unusual combination of qualifications as an accountant and as an electronics engineer. He was to play an important role in the financial team, together with C. W. ('Bill') Foreman, who had been recruited a few months earlier as Group Financial Accountant, from the British Printing Corporation, and was destined to succeed Denis Groom as Director of Finance.

Though Denis Groom found himself plunged almost immediately into the complexities of selling Vickers-Zimmer AG to Davy-Ashmore Limited, he and his young team also applied their minds urgently to the task of ensuring that financial information came to the centre quickly and regularly, that it was accurate information, and that it was effectively monitored. On paper the existing system seemed basically sound. Unfortunately, it required no more than quarterly reporting, and in practice, through slippages of one kind and another, a quarterly report could be as much as an additional two months behind events. The Board was thus too often receiving information which was both dated and misleading. Nor did the existing system call for regular reports on cash flow. Given their historical backgrounds, many operating units had their own banking arrangements with the local High Street bank, and at any given moment Vickers House could not be certain that it knew precisely the overall Group situation. In conditions of developing inflation this represented no small hazard, especially when the Rolls-Royce collapse towards the end of 1970 set rumours galloping about the liquidity position of other great firms, with Vickers specifically mentioned.

To remedy the internal reporting situation operating units were now required to follow up their approved annual management plan with promptly delivered monthly reports on performance under prescribed heads, and also to submit monthly cash reports and forecasts. A further decision of cardinal importance was to centralise banking arrangements as quickly as possible so that Vickers House could have precise knowledge and control of the cash and borrowing situation at all times.

To ensure an effective reporting and control system was one thing, but to deal with a deteriorating liquidity situation was another. Though this was a problem not exclusive to Vickers, the position in Vickers at the end of 1970 was disturbing, if not dangerous, and there were the additional uncertainties about the Zimmer business. For the Group as a whole short-term debt stood at £31m. Bank overdrafts totalled £16m and loans due for repayment in the next five years £15m, of which £10m were due in 1973 and 1974. Additionally, long-term debt stood at £15m to bring total borrowings to £45m. In relation to the size of the Group, and to Reserves which had now fallen to £28m, this was gearing too high for comfort in an inflationary situation. Nor were matters being helped by a high state of nervousness in the City, engendered by the position generally and the Rolls-Royce collapse in particular. With the need towards the end of 1970 to renew acceptance credits totalling £4m, Peter Matthews, accompanied by C. W. Foreman, found himself required to appear before a syndicate of merchant bankers. He made a sufficiently convincing impression for the credit to be renewed, but it was a disagreeable experience and left a resolve to repay the money as quickly as possible and not to seek further help of this kind.

Against this background it was clear that action must be taken urgently to improve the liquidity position. To assist him in dealing with the situation Peter Matthews moved J. H. Robbie from his executive chairmanship of the Engineering Group (though he remained non-executive chairman) so that Robbie's financial experience and expertise could be concentrated on this task. In essence it involved a drive to reduce stocks, which in many units had been allowed to accumulate unduly, measures to conserve cash, reduction in overheads, a drive on debtors and care in drawing contracts.

With so many contracts on a long-term basis, it was highly important to include arrangements for cash payments during the period of the contract. Too often, in order to win orders in depressed markets, contracts had been taken on onerous terms, without adequate cash payments, on over-strict specifications allied with penalties and with low profit margins. Because of the importance both of handling existing contracts which had run into difficulties and of ensuring that new contracts were not taken on adverse terms, Peter Matthews gave oversight of the Legal and Contracts Departments at Vickers House to W. C. P. McKie. Though he had joined Vickers a year before Peter Matthews appeared on the scene, Colin McKie had arrived by much the same route—Stewarts & Lloyds and British Steel Corporation—and the two men knew each other well. A lawyer by profession and equipped not only with the natural shrewdness of the Scots, but also with a razor-sharp mind, he believed, like Mrs Battle, in the 'rigour of the game' and eschewed sentiment in business. He had joined Vickers as Director of Manpower Development following the death of T. S. J. Anderson, and he retained a range of personnel responsibilities in addition to those he now assumed for legal and contract matters.

The Matthews team at the centre thus included a strong infusion of youth—Groom 38, McKie 38, Foreman 34, Butler 33—all recruited from outside the Company. A Director of Planning was also recruited from outside. This was B. T. Price, a physicist who had won a considerable reputation as a Civil Service 'boffin', principally at Harwell and subsequently in the Ministry of Defence and as Chief Scientific Adviser, Ministry of Transport. In the event, he did not find Vickers a congenial setting, not least perhaps because the degree of autonomy resting with the operating units made life frustrating for a planner at the centre. He was to remain in post for two years, and the appointment was then allowed to lapse, though a small planning department continued to operate in Vickers House with R. D. Gibbons as its manager.

If new blood had been introduced at Vickers House, senior executives in the traditional Vickers mould still played a powerful role. Leonard Redshaw continued to rule over the Shipbuilders and to hold the appointment of joint Assistant Managing Director of Vickers Limited. The other joint Assistant Managing Director, Colonel Jelf, also remained in post with chairmanship of the Printing Machinery and Supplies Group as his main executive responsibility.

To succeed J. H. Robbie as chief executive of the Engineering Group, Peter Matthews selected J. R. Hendin, then managing director of the South Marston Works. This was to prove an inspired appointment. In steering the Engineering Group to strength and stability – a sufficient achievement in itself – Jim Hendin was also helping to put new heart into the Vickers organisation as a whole and to establish greater freedom of manœuvre at the centre. A second-generation Vickers man – his father, a naval architect, had been General Manager at Naval Yard, Newcastle – he went to school at Rossall and then served a Vickers apprenticeship at Newcastle, gaining experience on the shop floor, not least in the heat and turmoil of the foundry. On moving into management he held a succession of senior production and sales posts in Newcastle and London before moving to South Marston Works in 1964 and becoming Managing Director there. His appointment as chief executive of the Engineering Group was very well received within the Group, where his cheerful pugnacity, total forthrightness, and capacity for immense hard work ensured that he was liked as well as respected.

He tells engagingly the story of his call to take over the Engineering Group. Following the tour of the works by Peter Matthews in the summer of 1970 speculation was rife about the changes that might take place when he formally moved into Vickers House on the 2nd September. Rumour said that he would be an axeman and had made it clear that no job should be considered safe. On the 2nd September J. R. Hendin was in Birmingham inspecting equipment installed at the

Post Office by South Marston Works when his secretary rang to say that Mr Matthews wished to dine with him privately the following night at the works guest house—a former farm house on the margins of the airfield. 'Oh well, this is it,' he told himself, 'the chopper for me.' Braced for the worst, he was stunned to find himself being invited, not to resign from running a works, but to take over leadership of the whole Engineering Group. There could be no illusions about the severity of the challenge, but events were to prove that he was the kind of man who thrived on power and responsibility.

By the end of 1970 the reorganised executive team at the top had thus already taken shape and Peter Matthews was clear about immediate objectives. Early in January 1971 he called a conference at Vickers House of senior managers from all the operating units in the United Kingdom. Disdaining oratorical flourishes, he set out the situation as he saw it and explained what he expected of them.

The change in direction which had taken place in the 1960s he described as 'decisive and bold and almost certainly right'. Nevertheless, urgent problems now existed, in part because of the violent inflationary situation. It followed that initially main emphasis must be given to the 'shorter term matters of profits and liquidity', not simply because of the need to deal with immediate difficulties, but also because 'without a satisfactory cash and profit position long-term planning becomes invalid'.

Action to deal with the situation, he went on, was essentially a matter of good day-to-day management—setting targets, checking progress regularly and frequently, and examining results self-critically.

'I belong to a school,' he said, 'who feel that in recent times Management Business Schools and other organisations are almost in danger of misleading able and intelligent young men into believing that the main issue in industry is decision making, whereas I believe that the main issue in large businesses lies in the implementation of decisions after they are made. This form of management is about making the best of people, about selling, controlling salesmen, about fixing prices and conditions of contracts of sale, about timekeeping, about yields, speeds, outputs and avoidance of waste, about the control of overheads, the control of stock and work-in-progress, about collecting debts, about reducing overdrafts, and above all, first and last, about influencing and controlling people.'

On prices and contracts, he emphasised that only in very exceptional cases should orders be taken without a reasonable level of profit. He did not like penalty clauses. He did not like orders which required major financing. He did not like 'those forms of risk where the reward is 10 per cent if everything goes perfectly and a loss in all other circumstances'.

On overheads, over-manning should not be regarded as acceptable, though he believed that it occurred less on the shop floor than in the office. There was no need to be ruthless or harsh, 'but it is worth reflecting that failure to grasp the nettle may not in the end be a kindness to anyone'.

On wages and salaries, 'I do not believe we can ever hope, nor should ever plan, to achieve higher profits by paying less than full price for labour or staff. In the long run, in this field as in most others, you get what you pay for.' He would, however, like to see 'wider differentials between the pay of the good, the mediocre and the poor performers. In certain areas Union and other activities preclude this. Let us not extend their thinking into the more senior fields of management. I want to see this Company as one which able men of all grades and classes wish to join because the Company is successful and because they can expect merit to be rewarded.'

Finally, he made it abundantly clear that he would fully exercise the 'wide executive powers' with which he had been entrusted and would fully discharge the responsibility he had been given 'under the authority of the Board for the initiation and formation of the commercial and financial policy of the Company'. Though it was his general intention to preserve the freedom and initiative of the Groups, 'I hope I have left you in no doubt whatsoever that I have a job to do, that I mean to see it done and that I require the services of the Head Office and of a satisfactory information system to achieve this.'

Organisationally, he further explained, because of the wide powers devolved on to himself as opposed to a committee of the Board, the Managing Director's Committee, though the centre of executive management of the business, would be advisory to him. With direct personal authority and authority delegated by him to the chairmen of the Groups it followed that boards and committees within the Groups would similarly be advisory to their chairmen.

As to the Management Committee, membership would comprise the heads of the main operating groups and the heads of four identified Head Office functions or groups of functions – finance, contracts and personnel, planning and public relations.

This Vickers House conference in January 1971 thus firmly charted the course immediately ahead. It also established the style of management favoured by the new chief executive – a style based, not remotely on paper and committees, but on close and regular personal contact by managers with the people they managed. For his own part, 'I expect and intend to spend at least one and generally two days a week away from Vickers House, meeting and talking with you in your factories and offices.'

In the meantime there was the question of adding up the score for 1970. From a poor result in 1966 pre-tax profit had advanced year by year to £8.4m in 1969. Trading profit in 1969, the annual report noted, was more than double that for 1967: subject to the usual reservations about trading conditions 'prospects were encouraging'. By September, however, when results for the first half of 1970 were announced, a much less sanguine note was being sounded. The damaging effect of inflation was now being felt, especially in the Engineering Group which found itself in mid-stream with a number of long-term contracts taken on terms not adequately anticipating the sharp rise in wages and the cost of raw materials. Inflation was also reflected in increased borrowing and therefore in higher interest charges. In fact, trading profit in 1970 would not have been greatly reduced but for the necessity to make provisions totalling £2.325m. This cut trading profit to £5.1m (£8.7m in 1969) and pre-tax profit to £4.2m (£8.4m in 1969).

If provisions were necessary this was the year to identify them. The market expected the new broom to sweep clean. If there were skeletons in the Vickers cupboards they wished to see them brought out. Much the biggest provision was for £1.55m in chemical engineering to meet losses expected on two contracts for the supply to the German Democratic Republic of a terephthalic acid plant and a polyester plant. A further £400,000 was provided for the loss expected by Medical Engineering on the Riyadh hospital contract in Saudi Arabia (£350,000 had already been provided). The Shipbuilding Group also held several contracts, including one for a cruise liner for a Danish consortium, on which provisions were deemed necessary to a total of £375,000.

In addition to these provisions, it was decided to write off against profits costs incurred by Medical Engineering in the development of their 'automated laboratory' – the Multichannel 300, a bio-chemical analyser – and of another piece of apparatus, the Vickers Auto-Tape System. The amount involved was £750,000 and this added to the provision for the Riyadh hospital meant that Medical Engineering showed a loss for the year of £1.2m. The other big loss-makers were the Engineering Group at £1.1m and Chemical Engineering £0.9m. Fortunately, Printing Machinery and Supplies had a good year, with a profit of £2.1m (chiefly from Howson-Algraphy), while Office Equipment and Supplies again turned in £2m. The Shipbuilders, though doing less well than in 1969, showed a trading profit of £1.7m and Australia and Canada brought their figure up to £1.2m from £0.9m in 1969.

When the Board met to decide dividend policy the figures on the table gave them little room for manœuvre. Profit before tax stood at only £4.21m and after tax £2.57m. After deduction of profit attributable to minority shareholders there remained £2.16m for appropriation. Since a 2½ per cent interim dividend had been declared in September at a cost of £1.09m virtually nothing remained for distribution. The 'traditional' Vickers 10 per cent for the year was obviously out of the question (a final dividend of 7½ per cent would cost £3.28m): in fact, any final dividend at all was virtually out of the question.

'In these circumstances,' said the Chairman's statement, 'and taking into account the prospect of continuing inflation in the economy, it is clearly of paramount importance to safeguard the liquidity of the Company. The Board therefore considered that there could be no justification for making any withdrawal from reserves in support of dividend payments and concluded with regret that they could not recommend payment of a final ordinary dividend.'

This was a traumatic moment. Never before had Vickers passed up a final dividend. The market did not react as pessimistically as might have been expected, partly because the possibility had been seen to exist, and partly because it felt that here was the new broom duly sweeping clean so that better things might be expected to lie ahead. Much depended on what the Chairman was prepared to put on record about future prospects. In his statement in the annual report, written in April, Sir Leslie felt it possible to say that '1971 should see the start of recovery from the severe setback in 1970'. Two months later, in his valedictory address at the annual general meeting, he was able to speak of action already put in hand to deal with the situation, including agreement in principle on the sale of Vickers-Zimmer AG to Davy-Ashmore Limited. He confirmed the expectation of 'a start to recovery'.

In referring to the sale of Vickers-Zimmer AG, Sir Leslie said that recent trends in the chemical engineering industry had indicated to the Board that a wider base was needed for the AG. In one particular area of chemical engineering it possessed a high degree of skill and know-how, but this now needed to be allied with activity in other sectors. To achieve this purpose the options were to put further investment into the business, to seek to arrange a merger, or to make an outright sale. In fact, the Board had accepted Peter Matthews' view that Vickers should cut its losses, actual and potential, as quickly as possible. There could be no question of further investment. A merger leaving Vickers as a lesser partner offered no attractions and the objective was therefore an outright sale. This depended, of course, on one of the major operators being sufficiently interested in the skills of Vickers-Zimmer to wish to acquire them. Here Vickers were fortunate to find Davy-Ashmore Limited poised to expand and wishing to do so particularly in West Germany and in the particular process technology offered by Zimmer.

Even so Vickers were negotiating from a position of weakness and had to make the best deal they could. Davy-Ashmore was interested only in acquiring Vickers-Zimmer AG, in Frankfurt, so that Vickers-Zimmer Limited in London and Compagnia Generale Resine, the Italian production and research facility, were excluded from the negotiations. Vickers-Zimmer Limited were formally contractors for the two plants in East Germany, for which provisions had been made in 1970, and Vickers had therefore to take responsibility for completion of these plants, though with Davy-Ashmore effectively its sub-contractors in doing so. Once these and some other outstanding commitments had been met Vickers-Zimmer Limited could cease to function. So, too, it would be necessary in due course to close or sell the Italian facility.

The purchase price for Vickers-Zimmer AG was agreed at £3.2m. This price was subject to amendment either way in the light of Zimmer's profit performance during the next five years and in the event, the Zimmer operation did well enough in its new setting for Vickers to receive additional payments.

The fact remained that withdrawal from chemical engineering proved a costly process. After tax relief of £900,000 the bill totalled £6.7m. This was no small sum to deduct from Reserves, already depleted (not least by Zimmer provisions), and by the end of 1971 the level of Reserves had sunk to £19m from £28m at the end of 1970.

Nevertheless, by getting chemical engineering out of its system, Vickers had freed itself of a business which it did not fully understand and could not properly control, and could now concentrate on building up its other activities. The loss to Reserves was heavy, but even by the end of 1971 the cash position had begun to look healthier, with bank overdrafts more than halved to £7m. The measures put in hand at the beginning of the year to deal with the inflationary situation had taken effect perhaps more quickly than could have been hoped.

2

Resurgence 1971–77

Having made its promised 'start to recovery' in 1971, Vickers was to produce better and better figures for the next five years until in 1976 its pre-tax profits stood at £38.3m, a spectacular advance even if subjected to inflation accounting. Success sometimes has a dull inevitability about it, but life with Vickers was never dull. Tracks may have been laid to the right destination, but there could be no guarantee that the train would get there smoothly. The economic signals could hold it up. The rolling stock was of different shapes and sizes and if the brakes jammed in one coach the whole train would be slowed down. Moreover, with bullion aboard, attacks by predators were more than likely. In fact, one sortie was quickly mounted, but the warriors of William Hudson melted away almost as silently as they had arrived. A second attack was altogether more serious and sustained since it came from Westminster itself with a concentrated assault on two of the coaches. That these two coaches would be detached and captured had to be accepted, and it was a matter of fighting a rearguard action while the others were strengthened and the train kept in motion. With these and other incidents the journey never lacked drama.

When it began in 1971 the City was prepared to suspend judgment about Vickers until the new regime had held office long enough for evidence to be available. This did not mean, however, that the City's antennae – particularly in the form of investment analysts – ceased to function. The Stock Market is never content to wait for published figures: what is likely to come must be foreseen and few modern stockbrokers fail to include in their armoury lively departments specialising in investment analysis and manned by polite, sharp-minded, sceptical young men, culled mostly from the universities. They work in part by analysis of published information, but also by visits to industrial companies to quiz the management. An industrial company does not *have* to talk to analysts and until the mid-sixties Vickers chose not to do so. After that the 'Welcome' mat was put out, reflecting the conclusion that whatever may have been the position when Vickers was primarily an armaments firm, clutching deadly secrets, the time had come for a full dialogue with the outside world. Given this conclusion, though one might talk also to City editors and their representatives, the investment analysts had a special role, not only in their own right but because they were also increasingly a source of information tapped by the financial writers. Obviously there were more links with the City grapevine than this, but the analysts were thought to be a sufficiently significant channel (with the bonus of incoming information) to warrant the considerable amount of time spent in talking to them.

There existed one short cut that could be taken from time to time —though only by invitation—and this took the form of a meeting with the assembled membership of the Society of Investment Analysts. It was, however, an occasion to be approached warily, to be undertaken only if you felt full confidence in the case you had to present, in your ability to make a good personal impression and in your ability to stand up to penetrative questions. In September 1972 Vickers had such an invitation and accepted it, with Lord Robens and Peter Matthews sharing the hot seat.

For Vickers Lord Robens was obviously a host in himself. It was clearly most unlikely that in 1971, with a whole range of offers before him on leaving the Coal Board, he would have attached his name and reputation to the Company without confidence in its future. When the announcement of his appointment was made the *Daily Telegraph* City column noted that Vickers shares rose to an extent sufficient to 'put back almost a million pounds to the market capitalisation of the Vickers equity'. A writer in the *Daily Mail* summed it up more colloquially. Lord Robens he described as 'the icing on the cake', adding: 'I reckon the shareholders will sleep easier tonight.'

When he addressed the Society of Investment Analysts Lord Robens was speaking against the background of unexciting figures for the first half of 1972. Though they showed a slightly better outcome than for the first half of 1971 this was more because of lower interest charges than improved trading figures. 'It has obviously been a difficult year,' he told the analysts. 'The anticipated improvement at the heavy end of engineering has simply not occurred and then, of course, there has been the succession of national upsets in industrial relations. However, it is fair to say that 1972 was in any case not the year when we expected to show a marked up-turn.' He also outlined the action taken to close, sell or re-align businesses for which a viable future could not be seen. 'In these processes,' he said, 'substantial reductions in manpower have been involved—in the order of 5,800, including 4,500 redundancies, in two years. The brunt of these reductions has fallen on the Engineering Group which now employs some 20 per cent less than it did in 1970.'

In summing up he declared his belief that Vickers was 'through the worst' and moving forward again in good order. 'Our shipbuilding activity is strong and stable. Out of the Engineering Group we aim to fashion a number of medium-size businesses with growth potential. We are likely to have an increasingly important presence in the office equipment and graphic arts industries. Those, then, are our main signposts.'

What could not be anticipated in September 1972 was the impact on Vickers affairs of three major developments – the so-called 'Rowland affair', the decision of the Labour Government to go ahead with nationalisation of both the shipbuilding and aircraft industries, and the remarkable 'take-off' in BAC profits.

'BAC', Lord Robens told the analysts, 'is in a pretty healthy condition and we regard it as a good investment.' Just how good was not immediately apparent, though BAC's own projections were encouraging. Indeed, at that time the Vickers Board was showing a good deal of reluctance to acquire, jointly with GEC, the 20 per cent holding in BAC which became available for purchase following the voluntary liquidation of Rolls-Royce. This was the original stake held in BAC by Bristol Aeroplane before it became a Rolls-Royce subsidiary, and the terms of the agreement under which BAC was set up required that any of the three partners wishing to withdraw must first offer its holding to the other two. Painful memories still existed in Vickers of the financial burdens of the Vanguard and VC10 and some directors saw an increase in Vickers holding in BAC as a reversal of the policy which had led to the 1960 merger. The eventual decision, in December 1972, to buy the extra 10 per cent, was therefore taken for defensive rather than positive reasons, since any other ownership, particularly by Hawker Siddeley, would almost certainly be to the disadvantage of Vickers. In fact, the cost at £2.8m (including a loan to BAC to replace half the loan previously made by Bristol) could not be regarded as exorbitant even at the time, but as things turned out it was to prove an extremely good 'buy' since it gave Vickers an extra 10 per cent share in the massive profits which BAC were to generate from 1973 onwards.

Whether profits would have been achieved on this scale had it been decided to go ahead with a project for a new airliner is a matter for speculation. The project came forward from BAC in 1969 and proposed a wide-cabin aircraft to be called the BAC Three-Eleven in succession to the One-Eleven. Without such a new venture, Sir George Edwards argued, BAC was in danger of becoming a producer only of military aircraft and guided weapons: they would have their factories at Weybridge, Hurn and Filton depending only on the last of the One-Elevens and the Concorde project, which was vulnerable to political cancellation. There would therefore be a risk to the large Vickers investment in BAC. On the other hand, for the project to be financially successful 240 aircraft would have to be sold on a time scale stretching into the 1980s.

Discussion on ways and means of finding the money for development and production of the Three-Eleven went on throughout 1970, the critical factor being the extent to which the Government would be prepared to provide launching aid. The position was reached when the Ministry indicated the likelihood that 60 per cent would be forthcoming provided the shareholders increased the equity in BAC from £40m to £50m. As the outcome of the only Vickers Board meeting ever to be held on a Sunday Vickers said reluctantly they would go along with their partners in finding this money in proportion to shareholdings – £4m each from Vickers and GEC and £2m from Rolls-Royce – by capitalisation of their existing loans to BAC. It was further agreed to promote a public issue of £12m unsecured loan stock, the expectation being that the City would cover £8m of this amount and GEC the remaining £4m. In fact, the offer produced City support only to the extent of £5m and Vickers then came under pressure to underwrite £2m: again, reluctantly, the Board agreed. At this point, however, in December 1970, the Government had a change of heart. It was no longer willing to provide launching aid. More than a year of discussion and negotiation had come to nothing, and it remained only for BAC to write off Three-Eleven development expenditure to date – estimated at £4m – and to declare redundancies, not least among their designers.

In the event it was almost certainly a blessing in disguise. BAC profits may have been pulled down in 1970, but thereafter, without the drain on resources that the Three-Eleven would have imposed, profits gained increasing momentum – £4.1m in 1971, £6.5m in 1972, £13.7m in 1973, £24.2m in 1974, £30m in 1975 and £40m in 1976. Cancellation of the Three-Eleven had certainly not constituted a risk to Vickers investment, whatever it may have meant in terms of reduced British presence in the business of designing and manufacturing commercial aircraft. The One-Eleven line lingered on and the Concorde project continued (for BAC more significant in providing work than profits), but what now really mattered to BAC was its central role in the military aircraft programme (especially Jaguar and Tornado), in guided weapons (especially Rapier) and to a lesser extent in space satellites and equipment. At the end of 1976 orders in hand stood at over £1,000m, of which three-quarters were for export. In particular, very large contracts had been won in Saudi Arabia and Iran.

With half its profits attributable to Vickers, BAC obviously made a major contribution to Vickers' own impressive results in 1974, 1975 and 1976. To this extent the new regime had its slice of good luck. Sir George Edwards saw the profits explosion in BAC as almost inevitable given that several major projects were simultaneously emerging from development into production. The extent to which the shareholders provided a stimulus could be argued. With Vickers and GEC in a 50:50 relationship, Peter Matthews felt that one of them should be recognisably in the driving seat and was content that this should be GEC. Under Sir Arnold Weinstock GEC had no doubts about the sharpness with which management should manage. He and his representatives brought considerable pressures on BAC and these efforts must have had their effect, whatever the inevitability of a profit explosion.

Though there could be nothing but satisfaction with the high-riding performance of BAC, Vickers could not be entirely happy that income from an associated company was outstripping trading profit from all its own operations. This situation might have been accepted more readily but for the fact that in 1973 it became clear that, should the Labour Party return to power, aircraft as well as shipbuilding would be on its shopping list for nationalisation. The possibility became a probability in 1974 following the General Election in February and the appointment of Mr Benn as Secretary of State for Industry. He made it immediately apparent that he proposed to launch a programme designed to bring the private sector of industry even more closely under State direction, including the establishment of a National Enterprise Board, a new Industry Act to give the Government wide powers to obtain information from companies, a system of 'planning agreements' with major companies, and nationalisation of the aircraft industry, together with shipbuilding, shiprepairing and marine engine building. With a further General Election in the autumn of 1974 giving Labour a small overall majority, the Queen's Speech at the beginning of the 1974–75 session of Parliament duly declared the Government's intention to introduce legislation and a Bill to nationalise both industries was laid before the House and at the end of April 1975. Thereafter the legislation followed a convoluted path, and towards the end of 1976 it even seemed possible that a General Election could intervene, but in the event the Bill became an Act, with the single casualty of shiprepairing, and vesting dates were set in April 1977 for the aircraft industry and in July for shipbuilding.

Thus for four years Vickers had been under the threat of losing not only its holding in BAC but also its shipbuilding activities, with all the uncertainties this entailed in forward planning. It did, however, provide an even sharper edge to the need to strengthen and expand the other segments of the Company's activities. The slow advance to nationalisation also gave more time to achieve this, but nice judgments had to be made about the extent to which the assumed inflow of compensation payments could be committed in advance to acquisitions and internal investment. Inevitably a programme of expansion would involve heavy borrowings, and this at a time when increased borrowing was likely in any event because of the pressures of inflation.

To replace the lost earnings of BAC and of the shipbuilders was obviously going to be extremely difficult and certainly could not be achieved quickly. With the shipbuilders also showing good profitability (£6.3m in 1974, £5.2m in 1975, and £4.7m in 1976), the combined contribution to Vickers pre-tax profits was running in the order of no less than 60 to 70 per cent. This was the size of the task. In absolute terms, however, encouragement lay in the steadily upward trend in the total of trading profits from the 'non-nationalisation' activities—£10.5m in 1973, £13.1m in 1974, £19.8m in 1975 and £22.6m in 1976. Within these totals the most satisfying feature was the advance of the Engineering Group in the UK from a loss position in 1972 to a trading profit of £6.2m in 1976. With the Australian and Canadian companies also on an upward trend engineering activities throughout the Group produced trading profits of £12.3m in 1975 and £12.1m in 1976. This was of the greatest importance in assessing the ability of Vickers to grow as a major international engineering company.

Looking then at the two other principal groups, office equipment and lithographic plates, progress during these years was also encouraging, though in the case of Roneo Vickers the advance was in turnover rather than profits. By 1976 Roneo Vickers had sales of no less than £113m and in this respect had become the biggest business in the Group. Its profits in the year, however, were £3.2m, a figure which had been matched in 1974 on a much smaller turnover of £71m. Nevertheless, reasons could be identified for this relatively poor profit performance, and in 1977 there was no lack of confidence in the Roneo Vickers management that the right actions had been taken and that sharply improving profitability could be expected. For Howson-Algraphy, however, it was in the nature of the business that the relationship of profit to turnover should be high, and though some vicissitudes were experienced in the five years 1972 to 1976 it was always a question of how high the annual profit would be. After four years in the range of £3m to £5m it advanced in 1976 to £6.8m, at which figure it was at the top of the Vickers profit league, followed by the Engineering Group at £6.2m and the Shipbuilding Group at £4.7m.

Among the smaller businesses Vickers Instruments showed an encouraging advance in profits in 1976, when nearly £400,000 was achieved. In search of expansion a 49 per cent interest was taken in Instruments' French agent, Etablissements Microscopes Nachet SA, itself a manufacturer of microscopes and accessories, and this was followed in 1977 by the acquisition of Joyce-Loebl Limited, based at Gateshead, a company specialising in the manufacture of advanced image analysis equipment. In 1977 John Munro, who had been managing director of Vickers Instruments for 13 years, retired though remaining a non-executive director: he was succeeded by S. S. L. Marshall, who had earlier been commercial director.

This general pattern of growing strength and improving profitability had been fashioned by a combination of new investment and tighter management controls over day-to-day operations. In his statement in the 1975 annual report, Lord Robens said that some £55m had been spent in the years 1972 to 1975 in re-investment. A year later he said: 'More has been achieved in this field in 1976 and a continuation of this vigorous policy of investment and expansion is planned for 1977 and 1978.' Adding in the cost of acquisitions and other investment undertaken earlier, a total of certainly not less than £100m had thus been spent in the years from 1965.

In the early stages of a recovery situation, a business must always regard itself as under scrutiny by those seeking opportunities for a cheap take-over, whether for asset-stripping, to acquire management ability or for perfectly good reasons of industrial logic. The pros and cons of taking over Vickers must have looked nicely balanced in the early 1970s. The share price was very low, but even so the amount required would put a deal beyond the reach of any organisation without large resources. Attractions included an assortment of valuable properties not essential to trading needs (not least the long-term lease of Millbank Tower) and the prospect of the inflow of a large amount of cash in compensation for the nationalisation of BAC and the shipbuilding group. Though the engineering businesses might present difficulties of disposal, Vickers also had a very large office equipment business which might in whole, or more probably in part, find keen buyers, and certainly Howson-Algraphy could be nothing but a glittering prize. On the other hand, as long as shipbuilding remained with Vickers, the Company's major role in defence contracting would surely signify Government opposition to a take-over by any organisation in which they did not have confidence.

Against this background, comments in the City pages about a strong rise in Vickers share price in the spring of 1973 began, towards the end of April, to associate the rise with buying attributed to David Rowland, described by *The Times* as 'the young financier, whose recent sale of a near 10 per cent share stake in the British Printing Corporation has helped to swell his cash resources'. *The Times* further spoke of leading institutions receiving bids for their Vickers stock and of deals being arranged above the market price. 'It is considered possible,' continued *The Times*, 'that Mr Rowland is attempting to line up a full scale take-over offer of Vickers. Backing for such a move would probably come in part from American sources. The attraction of Vickers is both its assets (the Millbank Tower) and an improving profits situation.' The *Daily Mail* said much the same in its more racy style. 'Young wheeler-dealer Mr David Rowland is said to be planning his biggest coup yet—a bid for Vickers, the armaments and engineering giant. Yesterday's hottest rumour was that Mr Rowland is in America gathering cash support to underwrite a bid for Vickers from Williams Hudson, the fuel distributors and wharf owners where Mr Rowland's Argo Caribbean offshore company own 43.5 per cent of the shares.' Other speculation doubted whether a bid was intended, especially given Vickers defence contracts. 'It looks like warehousing,' said the *Sunday Telegraph*, 'with Rowland happy to let the market know.' The theory behind letting the market know was that speculators would then rush in to buy, so forcing up the market price and making it possible for Rowland to sell all or part of his holding at a good profit: to the extent that if he didn't sell he would still have a holding likely to rise in value assuming that Vickers was indeed on a recovery course.

For the moment there was nothing that Vickers could do except read the speculation and make such enquiries as they could about David Rowland and Williams Hudson. Stock Exchange regulations did not require disclosure of a holding until it had reached 10 per cent and buying below this level could be hidden by use of nominee companies. Nor could share movements be immediately detected because the Share Register was normally six to eight weeks behind transactions. Confirmation of the reports at last came on the 9th May when a letter was delivered to Vickers House by hand from Williams Hudson Group Limited: it said that on the 26th April their holding in Vickers Ordinary Stock first exceeded 10 per cent and that by the

8th May the holding had reached £7,747,500, equal to 17.1 per cent of issue. On receiving this letter Peter Matthews rang the chief executive of Williams Hudson, Michael Kettle, to invite him to come to Vickers House to discuss the situation. The invitation was accepted, but a meeting did not take place until the 23rd May. Mr Kettle said that the buying had taken place because Vickers was recognised as a recovery stock and so as a good investment: Williams Hudson had been buying for some time and more recently had accelerated this buying, first in anticipation of the publication of good results for 1972 and then again after the results were announced. He did not, however, disclose future intentions – whether to go on buying, to bid or to sell. He made it clear that direct contact with David Rowland could not be expected, offered no criticism of Vickers management, and said that he hoped Vickers would not undertake any operation which diluted the equity, for example by merger with some other large company.

For Vickers an immediate need was to provide information and guidance to its employees, and a telex message was sent accordingly to all managements. It said that Williams Hudson Group had now become much the biggest single holder of Vickers equity. The Williams Hudson Group was thought to be under the control of Mr David Rowland, though he was not named as a director. The Argo Caribbean Group, registered in the Bahamas, held 43.5 per cent of the ordinary shares of William Hudson, whose business was described as 'transport and warehousing, fuel distribution, shipping and land and property development'. 'It is clear,' continued the telex, 'that the taking of a holding as large as this creates speculation about the possibility of a bid for Vickers. No bid has been made and it can be firmly stated that the Directors would vigorously oppose a bid which involved an association with a company which is not primarily engaged in manufacturing and is predominantly foreign-owned. In assessing any developments the Directors will have particular regard to the importance of maintaining Vickers as a major manufacturing company of national importance. They will also have strongly in mind their duties and responsibilities to all who work for Vickers as well as to the Company's stock-holders.'

In the event, the only development during succeeding months was further buying by Williams Hudson, duly notified, and by November the holding was well over 20 per cent. Motives remained undisclosed, with the *Guardian* City comment noting that Mr Rowland was residing 'virtually incommunicado in Paris'. In this state of uncertainty continuing speculation was inevitable. The *Guardian* reported signs of increasing Government concern. 'Vickers is after all the largest supplier of ships to the Royal Navy, and is the sole builder of Britain's nuclear-propelled submarine fleet. In addition Vickers acts as an overflow for much of the work which the Ministry of Defence's own ordnance factories cannot deal with themselves. A large chunk of the Iranian order for Chieftain tanks has gone to the group so that the order can be fulfilled within the specified time. Not only has Mr Rowland's reputation as an asset-stripper and profiteer obviously been fully noted by the defence top brass, but also there is a distinct flavour of "foreign interest" about Rowland's unwelcome holding.' The *Guardian* thought it probable that Rowland had approached a few large industrial groups, adding that GEC was 'thought to have given Rowland the cold shoulder'. While a bid from Rowland looked unlikely, 'if only because it must be beyond his resources, it would be idle to dismiss the fact that he is warehousing for somebody else'.

'Within Vickers,' added the *Guardian*, 'the attitude could best be described as one of uneasy calm.' In fact, within Vickers it was felt that Williams Hudson had bought to the point when they were almost certainly locked-in and would have to be rescued in due course by selling arranged in co-operation with Vickers and the City institutions. This proved to be the case. The moment of truth came on the 16th August 1974, with an announcement by Lazards and Morgan Grenfell that they had purchased from the Williams Hudson Group 10,360,000 ordinary stock units of £1 each of Vickers Limited, amounting to 23.68 per cent of the issued ordinary stock, and placed it with institutions at a price of 87.25p (subject to stamp duty). The announcement continued: 'The Board of Williams Hudson Group Limited considers that in the current financial climate the safeguarding of cash resources is and will continue to be of paramount importance. The sale of the stock provides the Group with a significant and immediate cash benefit and results in a substantial debt reduction, which itself gives a continuing cash flow advantage in the form of reduced interest charges. The Board of Williams Hudson Group considers that these benefits considerably outweigh the capital loss involved in the sale, which the Group is well able to absorb.'

The placing of the 10,360,000 units was completed within half an hour to 51 buyers, and for Vickers there could have been no better vote of confidence from the City. The capital loss sustained by Williams Hudson was variously estimated, with Lex in the *Financial Times* putting it at '£3m or more'. To have bought so large a stake, in the later stages at a relatively high price, was an extraordinary miscalculation by anyone as reputedly astute as David Rowland. It can only be assumed that whatever ploy he had in mind proved unworkable and that he was left in an untenable position, stranded high and dry perhaps by the cash flow problem of the mid-seventies.

The insecurity of life with Rowland did not inhibit Vickers from continuing effort to strengthen organisation and performance. Developments between 1972 and 1976 left few activities unaffected. Howson-Algraphy became an operating group in its own right, with printing machinery made a division of the Engineering Group. The acquisition of Dawson & Barfos Manufacturing Limited in 1973 made Vickers the largest British producer of bottling machinery. Table-top offset duplicators were added to the range of Roneo Vickers products by acquisition of a company in Federal Germany, Mathias Bauerle, under American ownership, at a cost of 6.5m US dollars. New factories were built at Crayford for the Engineering Group, in Holland and Spain for Howson-Algraphy (who also built a new research block and extended its Leeds works), at Scisset and Romford for Roneo Vickers and at Newcastle for Mitchell Bearings. A half interest was bought in the International Research and Development Company at Newcastle and this made possible the closure of Vickers Research Establishment at Sunninghill and sale of the premises for £750,000. A company was formed to develop properties in the Group not essential to trading activities. Further investment took place in Vickers Oceanics and the company was included with Brown Brothers and Slingsby Sailplanes in a new Offshore Engineering Group. The Vickers holding in Canadian Vickers was increased to some 72 per cent, and several acquisitions in Australia served not least to give the Vickers Australian companies an interest in the supply of equipment for rail transport.

Nor did the threatened 'State grab' of aircraft and shipbuilding inhibit Vickers from readiness to co-operate with the Government in projects in which clear industrial logic could be identified and to which no strings were attached. In 1974 the Department of Industry, deeply worried by the precarious state of Kearney & Trecker Marwin Limited, a leading British manufacturer of machine tools, approached Vickers with a proposition whereby the Department would inject new funds into the business provided Vickers undertook its management. As part of the arrangement Vickers were given an option to take over the company in 1976. In reorganising KTM Vickers concluded that its business and products could make a useful addition to engineering activities, and in 1976 a controlling interest was taken in the company. The Engineering Group thus acquired a new and sizeable activity with considerable potential. Vickers also co-operated with the Department of Industry in a scheme to restructure the business of John Hastie & Company, a Scottish firm manufacturing ships' steering gear, which had gone into receivership: again Vickers acquired the business, persuaded by the logic of adding Hastie's range of products to those of Brown Brothers in Edinburgh, and Hastie accordingly joined Browns in the Offshore Engineering Group.

While these events proceeded the operating groups were winning a number of prestigious successes. The Queen's Award to Industry was won twice by Howson-Algraphy and also by the Shipbuilding Group for export performance, and by Vickers Oceanics for technical achievement. Vickers Instruments won two Awards by the Council of Industrial Design for microscope design, particularly pleasing successes for one of the smaller Vickers businesses which rarely attracted attention precisely because without fuss it went on designing and making good products at a steady rate of profit. COID Awards were also won by Roneo Neopost, part of the office equipment group, for its franking machines, and by Brown Brothers for a stabiliser control unit.

Throughout this period up to 1977 Lord Robens remained non-executive Chairman and Peter Matthews chief executive, but a good deal of change took place in Board membership. The stature of Peter Matthews was now widely recognised, in industry, in the City and in Whitehall, and he came under pressure to serve on national councils and committees: appointments he accepted included membership of the National Research Development Council, the British Overseas Trade Board and the Export Guarantees Advisory Council. He also became a member of the Council of the CBI and chairman of its Research and Technology Committee. So, too, invitations came to join the boards of companies in the private sector and he was appointed a director of Lloyds Bank and British Electric Traction. In 1977 he accepted an invitation to become a non-executive director of Pegler-Hattersley Limited. Attempts were made by the Government to lure him into the public sector, but these he resisted though he did agree to become a non-executive Deputy Chairman of BSC and continued in this role until 1977. His knighthood in 1975 came as no surprise and was widely applauded.

Two long-serving non-executive members of the Board relinquished their appointments on reaching the age of 70 – Colonel Maxwell in 1975, after 41 years service, and D. L. Pollock in 1976, after 22 years. R. P. H. Yapp, an executive turned non-executive, also retired in 1976, having been a Vickers man (like his father) for 45 years and a member of the Board for 18. Among what could be called the old guard of non-executive directors there remained at the end of 1976 only Eric Faulkner, now knighted and chairman of Lloyds Bank. The other

13

non-executives at the end of 1976 were J. Martin Ritchie (appointed in 1970 when he was chairman of the Bowater Paper Corporation), F. R. P. Vinter (appointed in 1974, having recently retired from the Civil Service as a Deputy Secretary in the Department of Trade and Industry) and D. A. S. Plastow (appointed in 1976 and managing director of Rolls-Royce Motor Holdings).

Departures among long-serving executive directors were J. H. Robbie and A. P. Wickens in 1972, Robert Wonfor and Colonel Jelf in 1973 and Sir Leonard Redshaw in 1976. Of Sir Leonard Lord Robens wrote: 'No executive can have served Vickers with greater distinction and dedication. Under his leadership Vickers Shipbuilders have enhanced their already high international reputation and he has established himself as a leading personality in the British shipbuilding industry.' This view was reinforced in 1977 when he was awarded the John Smeaton medal of the Council of Engineering Institutions in recognition of his 'exceptional leadership and initiative in the development of advanced techniques in welded ship constructions in general and in submarine construction in particular'. It was not, however, a final farewell since Sir Leonard had agreed to act as non-executive chairman of the Offshore Engineering Group 'in whose formation and development he has played a key role'. His successor as shipbuilding chairman was William Richardson.

Newcomers as executive directors were D. J. Groom in 1971, J. R. Hendin and W. C. P. McKie in 1972, C. W. Foreman in 1973, K. W. Ketteringham in 1975 and R. O. Taylor in 1977. For the first time the chief executives of Canadian Vickers and Vickers Australia also became members—J. E. Harrington in 1974 and P. D. Scott Maxwell in 1975—though in view of the long distance from London of their operating bases it was not expected that they would attend Board meetings regularly.

Denis Groom served as Director of Finance for a little over two years before moving into the chairmanship of Roneo Vickers. He had held this responsibility for little more than a year when he was offered an important appointment in Canada and found the offer too tempting to resist, especially against the background of Britain's punitive tax system. He left the Company in June 1974. His successor as Roneo Vickers chief executive was K. W. Ketteringham which meant that after a number of false starts the office equipment group was in charge of an executive from the Roneo side of the organisation. Cheerfully extrovert and a man of immense energy and determination, he knew the business intimately from 20 years experience in a succession of senior appointments. He had very clear and firm ideas about the changes needed in the Roneo structure and proceeded to carry them out, supported by an able and enthusiastic top management team. He was invited to join Vickers Limited Board a year after taking over at Croydon.

At the time of his appointment to the Board Colin McKie had been with the company for three years and was still only 40. His original appointment had been as Director of Manpower Development, but he had quickly decided that responsibility for training, including sponsorship on degree courses, should properly rest with the operating groups, and that an education department at Vickers House could not be justified. The department had accordingly been dismantled in 1971, and with it the scheme for an annual award of four university scholarships to sixth formers. During the ten years in which the scheme was operated 43 awards had been made at a cost of some £130,000. With no strings attached to the scholarships, 19 of the 43 holders had not joined Vickers on getting degrees. The remaining 24 took up Vickers appointments, but by 1977 only seven were with the company, still young men, but all in senior appointments, chiefly in the Engineering Group. Having demolished most of this range of his responsibilities, McKie had taken on others, and was Director of Contracts when he joined the Board: in this role he tackled not least a number of situations in which major contracts had turned sour. Other subsequent responsibilities included chairmanship of the properties company and of Vickers Instruments. He also steered a new Group pension scheme into existence and became chairman of Vickers Pension Trustees Limited. Yet another task was to head an executive committee formed to direct and co-ordinate action and negotiations on the nationalisation issue.

The appointment of C. W. Foreman as Director of Finance in succession to Denis Groom brought an even younger man into the Board room. He was then only 36. Had there been any misgivings about a man as young as this joining the top reaches of management they were quickly dispelled. His calm and articulate professional competence, allied with a relaxed friendliness, ensured his acceptability not only at the Board table, but also in the offices of the operating units, not least at Barrow. Perhaps in the Vickers setting it was a further advantage that he had northern origins and could list the Royal Grammar School, Newcastle, among his places of education.

Under the Matthews regime annual management conferences continued to provide a forum for reviewing progress and charting the way ahead. Usually they were held at centres near to the main Vickers establishments. In 1977 the venue was Brighton, home of the recently acquired Kearney & Trecker Marwin. This was the first conference following the Act to nationalise shipbuilding and aircraft, and so took place without familiar faces from Barrow. To that extent it was an occasion for regret. An observer would have had difficulty, however, in detecting any mood other than optimism and confidence. Possibly the significance of the loss of shipbuilding and the investment in BAC had not yet been fully assimilated by many of those present. Though replacement of the lost profits would clearly be difficult, the fact remained that trading profit from the ongoing businesses had increased from £13.3m in 1974 to £22.6m in 1976 and was expected to maintain an upward trend, subject obviously to whatever might happen in the economy and in industrial relations. In part the upward trend reflected investment and acquisitions undertaken ahead of receipt of compensation monies for nationalisation, as could be seen from a borrowings figure of £86m at the end of 1976 compared with £22m at the end of 1973. On the other hand, Reserves had grown from £73m to £112m during the same three years, resulting in part from revaluation of assets in 1973, but including £30m in retained profits. Earnings per share had doubled from 20p in 1974 to over 40p in 1976. No doubt some stockholders would have considered it right and proper for more to have been distributed in dividends and less retained, but even had the Board wished to do this their freedom of action was limited by statutory dividend controls, and by the fact that part of the increased earnings were retained in BAC. Even so the dividend had been increased each year from the very low figure of 2½p per pound gross in 1970 to 13.5178p in 1976.

The management conference at Brighton demonstrated that in mid-1977 internal confidence in the future of the Company stood high. Externally it was shown in a share price rising to over 200.

3

The Engineers

When J. R. Hendin became chief executive of the Engineering Group in September 1970 he knew that the years immediately ahead could only be difficult and disagreeable. The Group was running at a loss. Some of the ailing businesses and products would have to be rooted out. Changes would have to be made at the highest levels of management and these would include men counted as personal friends. Perhaps it was as well that the crystal ball did not also fully reveal how hostile the economic and industrial climate would become. It could be counted to advantage, however, that no overnight miracles were expected. Given that large-scale redundancies would be necessary, in addition to natural wastage, it was also to advantage that statutory payments reduced the hardship, to the extent indeed that in most situations voluntary redundancy of well over 50 per cent could be expected. In dealing with these situations, the new chief executive made a point of appearing personally to explain reasons and to discuss ways and means of minimising dislocation.

Figures indicate the degree of reorganisation. During the years 1970 to 1976 the Engineering Group declared a total of 3,500 redundancies of which 2,700 took place in the three years 1971, 1972 and 1973. In terms of numbers employed, however, total strength of the Group at the end of 1976 was only 400 less than at the end of 1970 – 8,900 against 9,300. The closures of the Boby Works, ABC Motors and the All Wheel Drive business, plus large-scale redundancies at Scotswood, South Marston and Crabtree-Vickers, had been offset by the acquisition of Dawson and Barfos (700 employees) and KTM (1,000) and the setting up of Design and Projects Division (265). Perhaps the most significant – and encouraging – feature of these changes lay in comparisons of turnover and productivity. With a reduced number of employees the Group's turnover at the end of 1976 stood at £78m, exactly double that in 1970, and turnover per employee was more than double at £8,710. Return on capital employed had gone from nil to 15.9 per cent.

Taking out losers, building up winners, getting the right people into the right jobs, and insistence on high management performance coupled with strict financial controls – all played their part in the Group's recovery. It was aided, too, by broad stability in industrial relations. Though difficult situations arose, and militancy sometimes flexed its muscles, major disruption was avoided by willingness on the part of management to table facts and figures, to explain and to discuss, and on the part of the unions to recognise where the limits should be drawn when the crunch came. The relative smallness of most of the engineering units no doubt also played its part. Few had more than 1,000 employees which meant that senior management was never far from the shop floor and that communication presented no intractable problems.

Though industrial relations was very much a local responsibility, Jim Hendin made sure that in this, as in other matters, there was no feeling of remoteness from the centre. On becoming chief executive he had removed engineering group headquarters from Vickers House to offices at South Marston Works. The airfield there made it possible for him to pay quick and frequent visits by Company aircraft to the northern works, and rail and motorway links put him in easy reach of London and Crayford.

The future of Crayford Works posed immediate problems.

The works had a colourful past, reaching back to 1884 when members of the Vickers family invested in a new company formed to manufacture the automatic gun invented by Hiram Maxim. It was from this weapon – the first truly automatic gun – that the Vickers machine gun was developed. To make the Maxim guns a factory was built at Crayford. In his autobiography, Maxim wrote: 'We looked about for a larger place, and finally found a lot of large, empty buildings at Crayford, in Kent, which were very suitable for the purpose. There was a large boiler house with two large Lancashire boilers, and the landlord would not let us take over the premises unless we purchased the boilers.' Maxim then described how the landlord tried to get him to pay £800 for the boilers, though on valuation by 'a real engineer' their value was put at £30 – 'just what they were worth in old iron'.

Maxim was also interested in the theories of heavier-than-air flight and to prove his theories built a flying machine, an extraordinary steam-driven contraption which might have served as inspiration for Heath Robinson. It did, in fact, become airborne in Baldwyn's Park, Bexley Heath, which Maxim had hired for the experiments, but its lift was deliberately limited by an overhead 'track'. Many distinguished visitors came to ride in the machine, including the future King George

V, who demanded 'Let her go for all she's worth'. On an occasion when 'one of our directors' came to see it, the engines were turned to full speed and the machine rose to such effect that, says Maxim, 'I found myself floating in the air with the feeling of being in a boat.' In doing so, unfortunately, it broke the upper track and the machine was badly damaged. 'This,' claimed Maxim, 'was the first time in the history of the world that a flying machine actually lifted itself and a man into the air.' The year was 1894.

Three years later Vickers acquired the Maxim Company, and the factory's activities were widened though gun manufacture continued. The works grew rapidly, multiplying still further on the outbreak of World War One when it manufactured not only guns but also aircraft, notably the 'gun-bus', armed with a Vickers gun and claimed to be the world's first real fighting aircraft. It was built in large numbers for the Royal Flying Corps, and was followed by the SE.5A single-seat fighter and by the first of the Vickers Vimy bombers. 'Crayford,' said the Works Magazine in 1917, 'is the only Works in the world that can turn out a complete aeroplane, with the exception of instruments such as compasses, but including the engine and armament of machine and rocket guns.' It also made over 50,000 Vickers machine guns, shells for artillery, mines and a variety of other requirements for military equipment. By the end of the war it was employing 12,000 people.

With peace came the scramble to find work through commercial products. These included sewing machines, washing machines, motor cycle engines, sporting guns and textile machinery. A new and important armament product was acquired in 1931 when rationalisation led to the closure of the Vickers Works at Erith and transfer to Crayford of its production of fire control equipment for naval and military use, together with boxmaking machinery and hardness testing machines. These three activities were to provide Crayford with a basic workload for many years to come. World War Two inevitably put Crayford back into the front-line of armament production and its workforce again grew to over 10,000. Among its products was the predictor, an instrument vital in anti-aircraft defence. Crayford also took part in manufacturing the bouncing bombs invented by Barnes Wallis.

Once again, after the war, Crayford had the difficult task of finding products to fill denuded capacity. Fortunately, it still had its established activities in fire-control equipment, boxmaking machinery and hardness testing machines, and to these were now added accounting machinery (achieved by purchase in 1945 of an interest in Powers-Samas Accounting Machines), petrol pumps (through an agreement in 1946 with Gilbert & Barker, an American company), printing machinery (offset litho presses from the overloaded order book of George Mann & Company, acquired by Vickers in 1947) and bottling and brewing machinery (through the acquisition, also in 1947, of G. J. Worssam & Son, manufacturers of filling machinery and associated brewing equipment). Though providing useful work for varying periods the manufacturing undertaken for Powers-Samas, Gilbert & Barker and George Mann fell away as those companies were reorganised, and by the mid-1960s Crayford was concentrating on fire-control equipment, bottling machinery, packaging machinery and hardness testing machines.

Cuts in defence spending were now bringing a lower level of orders for fire-control equipment, and this was a factor leading to a scheme to reorganise the works early in 1968. Any such scheme was bound to aim at more economic use of the extensive land and buildings occupied by the works, and the upshot was a reduction of about one-third of the floor area in use coupled with release of a substantial amount of property. Some 200 redundancies resulted and by the end of 1968 the workforce was reduced from 1,800 to 1,400. Most of the site was then sold to Rich Estates Limited for £2m with arrangements to lease back for five years the area required by the works activities. During this breathing space the longer-term future of the works would have to be determined by the light of performance in the new setting.

This was the position in 1970 when Jim Hendin took over the Engineering Group. Basically it had to be decided whether to close down at Crayford completely, moving the viable activities to South Marston, or to build a new factory on the land retained at the time of the sale to Rich Estates. The policy of streamlining engineering activities extended beyond Crayford, of course, and two businesses whose continuing viability as separate works stood in doubt were those of ABC Motors at Walton-on-Thames and Robert Boby at Bury St Edmunds. Early in 1971 the ABC Motors business was moved to Crayford and the premises at Walton-on-Thames sold. Some key personnel transferred to Crayford, but 200 redundancies were involved. ABC products included the Vickers Inch/Metric conversion drum, for which the impending change to the metric system offered promising prospects. Responsibility for the Boby business was placed with the Bottling Division at Crayford, but this arrangement was relatively short-lived as the Boby Works was sold in 1971, and the business placed on a design and procurement basis operating from South Marston. In the meantime, efforts had been made to build up bottling machinery activities, notably by the acquisition early in 1971 of a Belgian firm, Usines

Arthur Vandergeeten, who in particular brought bottle washing machinery into the Vickers range. So, too, an acquisition was made in order to strengthen the range of products of the Container and Packaging Machinery Division: this was Kirby (Engineers) Limited at Walsall.*

Despite these developments the future of activity on the Crayford site continued in doubt, especially when it became clear that orders from the Ministry of Defence for fire-control equipment were likely to tail off. This work occupied some 25 to 30 per cent of Crayford's workload and its cessation would obviously pose a threat to the whole Crayford operation. In November 1971 the workforce was warned accordingly that defence work was to be phased out and an examination made of Crayford's other activities with a view to consolidating them into a still more compact organisation. The prospect greatly alarmed both trade unions and local authorities and they joined with the local Members of Parliament to make representations to the Ministry of Defence about the continuation of armament orders. These efforts had partial success and Vickers announced in May 1972 that it was investigating the possibility of building a new factory on that portion of the site retained by the Company: this was expected to be a smaller works so that substantial redundancies would be involved even if the scheme went forward. Local alarm was thus not completely abated and a meeting took place in July with representatives of the Bexley Council and District Trade Union officials at which Jim Hendin explained the situation as Vickers saw it. To an extent this meeting cleared the air and Vickers now began to examine proposals for further expansions of the bottling machinery business. In November it was also announced that armament work would continue for at least another two years. With the situation thus held, Vickers took a further look at the various possibilities, and concluded that the prospects in the commercial divisions at Crayford, particularly in bottling machinery, justified going forward with the building of a new factory larger than originally envisaged and involving no further redundancies.

So Crayford had its new factory, built at a cost of £2m and opened by Lord Robens on the 27th October 1975. By then major developments had taken place in the bottling machinery business. One was highly disappointing—the performance of Vandergeeten, the Belgian company acquired in 1971. It had consistently traded at a loss and was basically insolvent. In 1975 Vickers decided that it could no longer put resources into Vandergeeten and the business accordingly went into liquidation. This setback was much more than offset, however, by the

*In the event this acquisition did not fulfil hopes and the business was sold in 1975.

acquisition towards the end of 1973 of Dawson & Barfos Manufacturing Limited. With the products thus acquired it could be claimed in Vickers 1975 annual report that Vickers bottling machinery business was 'now much the biggest of its kind in the UK, with the ability to offer a complete range of fillers and washers for the brewing, soft drinks and dairy industries'. The Dawson & Barfos factories at Gomersal (Leeds) and Thetford employed some 700 people, about the same number as at Crayford, and the new Vickers-Dawson Division, with headquarters at Crayford, had a total strength in 1976 of nearly 1,250 with turnover approaching £10m and pre-tax profit approaching £1m. It had thus become one of the main businesses in the Engineering Group. On turnover, Defence Systems, as the Armament Division had now been re-named, was still the biggest business, followed by Vickers-Dawson and Crabtree-Vickers at about the same levels with Design & Projects coming up strongly.

While the Crayford saga unfolded, other problem situations had also to be unravelled. Printing machinery, hydraulics and medical engineering had all failed to fulfil the high expectations placed on them.

From the status of a group in its own right, Printing Machinery and Supplies was divided in 1972 into two main components – Crabtree-Vickers, concerned with printing machinery, and Howson-Algraphy, concerned with printing plates, chemicals and associated equipment. Crabtree-Vickers became a division of the Engineering Group and Howson-Algraphy a separate business, later to be given group status, but for the time being placed under the direction of Robert Wonfor, chairman of Roneo Vickers. With the Howson-Algraphy component removed it was possible to identify more precisely the performance of the three elements in printing machinery. The business in metal decorating presses was clearly the most successful, with products of high reputation in generally good demand and results measured from year to year in degree of profitability. The sheet-fed business, if less consistently profitable, seemed basically sound, but the newspaper press business was another matter. To equip a newspaper with printing machines represented a very large contract, usually with special features, and once the machines had been installed they were likely to be there for a very long time. In conditions of world trade recession, newspapers would be loath to embark on the major expenditure involved in new presses, and to this extent the manufacturers suffered from the high quality and long life of their products. For the manufacturer it was also a highly competitive business so that in times of slump contracts could be taken which were too onerous in their specifications on time and performance and too restricted in their margins. Crabtree-Vickers now found themselves both short of orders, or prospect of orders, and saddled with two unremunerative contracts. In this situation J. R. Hendin recommended to the Board that losses should be cut and the manufacture of newspaper presses discontinued. His recommendation was accepted and the business was then reorganised. The head office organisation at Leeds was disbanded, with a sharp reduction in overhead costs, and the sheet-fed and metal decorating businesses were required to stand on their own feet, which they willingly did. Even with the excision of newspaper presses, printing machinery remained one of the largest and most important activities in the Engineering Group, with some 1,300 employees, turnover in the order of £10m and a fluctuating level of profit.

On hydraulics, Vickers' perpetual problem child, the balance between closure and continuation was finely balanced for Hendin as for his predecessors. He believed, however, that it was a business in which, however cyclical, Vickers ought to succeed, and the Board agreed to proposals for reorganisation and further investment, backed by an interest relief grant from the Department of Industry. To some extent the Division's problems were associated with lack of new products and a major research and development project on high pressure pumps had been put in hand earlier at Vickers Research Establishment before its closure. The project manager, Peter Price, continued the work at IRD, in Newcastle, in association with the Division at South Marston, and the outcome was a new range of pumps, christened Dynapad, which were expected to put the business into an altogether stronger position.

In Medical Engineering it had to be accepted that the Multichannel 300 biochemical analyser could not be sold profitably, however efficiently it might now perform, and also that the business had expanded too ambitiously. Withdrawal from the activity in hospital gas pipelines (by sale to the British Oxygen Company) and from active marketing of the Multi channel in the USA, led to a reorganisation in 1975 involving concentration on paediatric and oxygen therapy equipment, in which a lively business had existed for some years, and on small analysers and improvements in laboratory data processing. In particular, a 'software' system for use with biochemical analysers had been successfully developed and was selling well. By 1976, with one of the Group's bright young men, C. J. O'Donnell, as its Managing Director, the business was trading profitably and thought could be given to ways and means of achieving growth.

Problems for the heavy engineers in Newcastle related principally to the Scotswood Works. Here the cyclical nature of the power press business, and the lack of other 'standard' products, except for Paceco cranes, and equipment for submarine cable telephone systems, left the works too vulnerable to fluctuations in jobbing engineering. Nevertheless, once a number of large but unremunerative orders had been worked out of the system, the works moved into profitability, and plans were laid for product diversification. At Elswick, the situation depended essentially on the state of the order book and prospects in Defence Systems Division. Though the traditional alarm signals rang from time to time, the Division worked mostly at full stretch, principally on overspill contracts from the ordnance factories. Hopes for the future were bolstered in 1977 by a new overseas order for Vickers' own main battle tank. It seemed likely, therefore, that with the Non-Ferrous Metal and Pressings Divisions continuing to do well, Elswick would stay on an even keel. The one activity in doubt was the Foundry which had been a constant loss-maker despite its modernisation and the business was sold in 1977.

The third Vickers manufacturing activity in Newcastle, the Michell Bearings Division, had come under full Vickers control at the beginning of 1969, when the Company bought for some £220,000 the 70 per cent of the shares held by other interests. Though its products were well established and of good repute, the business at first had a mediocre profit performance, not least because of under-charging. With the introduction of new senior management in 1972, however, the situation changed almost dramatically and from 1973 onwards it enjoyed increasing success, with good profit from good order books and impressive export figures. With performance and prospects so encouraging the decision was taken in 1976 to build a completely new factory at a cost of £4.5m gross. The existing site was large enough for this to be undertaken by stages, without undue dislocation, and the first phase was completed on schedule in April 1977.

Vickers thus had a very strong presence in Newcastle, and the doubts and uncertainties of previous years had been largely dispelled. In the process of reorganisation, the slimming down at Elswick and Scotswood had released land for development by the property company and its associates, without inhibiting manufacturing operations. Moreover, Vickers had added to its interests in Newcastle, first by becoming joint owners of International Research and Development, and then in 1975 by building premises designed to provide a computer centre serving not only the Engineering Group but also Vickers House, Howson-Algraphy, Vickers Oceanics and Roneo Vickers Business Forms. Interlink House was opened by Sir Peter Matthews in October 1975. Using systems developed in conjunction with ICL and the Department of

Industry, it had sufficient capacity to provide services to organisations outside the Vickers Group.

In addition to the emergence of Michell Bearings and Vickers-Dawson as major businesses, the 1970s saw a rapid advance by Design and Projects Division and an extension into the machine tools industry by acquisition of control of Kearney & Trecker Marwin Limited.

From the time of Vickers agreement in 1974 with the Department of Industry to undertake management responsibility for KTM a considerable amount of senior management time was taken up in its affairs, particularly that of J. R. Hendin and C. W. Foreman, and doubts inevitably existed about its justification. In fact, a thorough examination of KTM financial and commercial systems established that improvements could be introduced which would have every chance of establishing KTM as a viable business. This examination was undertaken mainly by Ray Enticott, a chartered accountant who had joined Vickers in 1958 and held senior appointments in Newcastle. Not least he had played a key role in the recovery and expansion of Michell Bearings as its commercial manager. Given confidence that KTM could be built into a profitable business, the legal and other complexities involved in its reorganisation by Vickers could be tackled without inner feelings of time wasted. Finally, in August 1976, after legal proceedings taken by one of the minority shareholders, it became possible for Vickers, with the approval of the Department of Industry, to acquire over 85 per cent of the voting share capital at a cost of £803,000. In 1977 the business produced a profit in the order of £1m, coupled with a strong cash flow. The chief executive was then Kenneth Lane, who had come to Vickers in 1975 after 20 years experience in the electrical engineering industry.

Meanwhile, the Design and Projects Division at South Marston was growing rapidly in stature. From a basic workload of main generator connections, it was proving its competence to undertake the design and installation of very large and sophisticated engineering projects. Its success in doing so brought a good deal of manufacturing work to the engineering group as a whole. An example of its achievements was the dynamometer, a machine for testing aircraft tyres, wheels and brakes. Developed first for Dunlop, the dynamometer was subsequently ordered by the USSR and by the Republic of China, both multi-million pound projects. Though in the nature of its activities not a large employer, the Division had by 1976 a highly skilled staff of over 250, and had established an ancillary office in a new building at Eastleigh, near Southampton.

With headquarters remaining at South Marston it was one of the activities occupying that part of the site now being developed, in effect, as a trading estate. The main Vickers activity at South Marston, in terms of numbers employed, continued to be Hydraulics Division with the small but successful Nuclear Engineering Division as a neighbour and also Automated Systems Division, hived off from the Roneo Vickers Group to carry forward work on mechanical filing systems and on materials handling installations. This was an activity stemming from the installation in the new headquarters of the Halifax Building Society of an automatic filing system claimed to be the biggest in the world.

In carrying out the reconstruction of the Engineering Group so successfully J. R. Hendin had the backing of an enthusiastic team of senior managers. He had made a point of giving several young men a taste of high responsibility and they responded well. One was W. M. Windsor, who had joined the Company at Vickers House in June 1971, becoming Deputy Manager, Group Project and Financial Evaluation Department, shortly afterwards. Previously he was Manager of the Contracts Department of GEC Elliot. He moved to Crayford Works two months later as Managing Director, Container and Packaging Division, before becoming Managing Director of Crayford Works in May 1973 at the age of 36. His appointment as Chairman and Managing Director of Vickers-Dawson followed in February 1974. This extra-ordinary rapid rise reflected his ability to take increasing responsibility in his stride and to discharge it with competence and confidence. Another young man coming to the fore in the Hendin team was J. M. Harper, appointed Managing Director, South Marston Works at the age of 30 in June 1975. Son of a Vickers man, Barrow-born and educated, he became a B.Sc. as a Vickers Scholar, and on joining the Company quickly rose to senior appointments, including that of production director at Crabtree-Vickers. Also representing the younger generation on the Engineering Group Board was C. V. Chester-Browne who was appointed Group Technical Director in January 1973 at the age of 39. His duties included oversight of development of management information and control systems, and he was also in charge of the burgeoning Design & Projects Division.

Two men were brought in to become chief executives of operating units. One was K. R. Hoskins, an Australian, who had served an engineering apprenticeship in the United Kingdom, and then worked in Australia and India before returning to Britain: he became managing director of Crabtree-Vickers in August 1972 and remained in the appointment until February 1977. The other, Peter M. Crowther, came from a senior post with Babcock & Wilcox to take charge of the Armament Division at Elswick in January 1975.

The old guard of Vickers managers remained strongly represented, however. Harold Rodger, another man from the Barrow environment, was a key figure at Engineering Group headquarters as Group Commercial Director from January 1970. His motto, as with Prime Minister Macmillan, might well have come from Gilbert and Sullivan: 'Cool, calm deliberation disentangles every knot'. A. W. Taylor, a Vickers man all his working life, beginning with an apprenticeship at Crayford, moved from chief executive, Crayford, to chief executive, Scotswood, in March 1972 and presided over Scotswood's improving fortunes. Another long-serving Vickers manager, W. Richardson, after fighting rearguard actions at Ioco and Bobys, went to Newcastle in November 1971 as chief executive of Michell Bearings, and before his retirement had the satisfaction of leading the business to much higher levels of performance, with prospects of even better things to come. In addition to Harold Rodger, the Engineering Group Board had two functional directors. These were Don Giles, Director of Marketing and Sales (previously he had been in charge of contracts and marketing at Scotswood), and L. C. Libby, formerly a professional soldier in the rank of brigadier, whose talent for understanding people led to his appointment as Director of Personnel and Manpower Development, after stints as chief executive at Medical Engineering, Crayford and South Marston.

In a reorganisation announced in October 1977 Harold Rodger assumed the role of Deputy Chairman, while continuing as Commercial Director, and two of the younger men, C. N. Davies and W. M. Windsor, came to the top as Joint Managing Directors, Davies operating from Newcastle and Windsor from Crayford. Noel Davies came to this appointment at the beginning of 1978 from Barrow Engineering Works where as General Manager he had been among those 'taken over' by British Shipbuilders when Vickers shipbuilding activities were nationalised in July. A Bachelor of Science and a chartered engineer, he had joined Vickers in 1956 after an apprenticeship with the Austin Motor Company. He became a member of the Vickers team at Harwell, and then at the nuclear submarine prototype at Dounereay, being chief engineer at Dounereay when the prototype was handed over to the Royal Navy.

4

More than Desks

Following the acquisition of Roneo Limited in 1966 and the formation of a Roneo Vickers Office Equipment Group, figures of sales and trading profit for the new Group were published separately for the first time in 1967. Sales then stood at £16.3m. In each of the nine subsequent years sales increased, at first steadily and then with increasing momentum by jumps of some £20m per year. In 1976 not only was the £100m mark reached, but the total went up to £113.7m. Clearly Roneo Vickers were very good at marketing and selling. To an extent the increase reflected acquisitions. Even so, by the yardstick of sales per employee, the figures were impressive – from £4,571 in 1971 to £11,488 in 1976. Additions to manpower through acquisitions had been largely offset by reductions totalling 1,200 in streamlining operations. In 1976 Roneo Vickers had 9,395 employees compared with 8,394 six years earlier in 1971.

On the face of it, a highly satisfactory growth in the business. It was another matter, however, if one looked at figures for profit and return on capital. Trading profit in 1971 on sales of £38.4m was £2m, certainly not an exciting figure, but giving a return on capital of 12.1 per cent. Against sales three times as great in 1976 profit had risen to only £3.2m and return on capital had fallen to 6.2 per cent. The best profit figure had been £3.8m in 1975, but it seemed that as turnover multiplied profit remained stagnant at an unacceptably low level. This was especially disappointing when it became clear in the mid-1970s that, following nationalisation of shipbuilding and BAC, Vickers would need to obtain much greater earnings from its ongoing businesses if the gap in profits was to be bridged. Most of the other businesses were on an upward trend, but a good deal had been expected from Roneo Vickers and the absence of profit growth at this juncture was bound to be disturbing. Certainly by 1976 it had been hoped that office equipment would be showing results justifying the substantial investment of recent years. If senior management at Croydon shared this disappointment there was nevertheless confidence in 1977 that for well identified reasons it was simply a matter of joy deferred: during the last few years it had been a case of keeping the shop open while rebuilding it: the new shop was now ready: very quickly profit would be at altogether higher levels.

Various views could be taken on the reasons for the ten-year phenomenon of soaring sales and stagnant profits. One school argued: 'They've been too busy making acquisitions to make profits.' Another: 'They're in too many activities and not big enough in any of them.' A third: 'Their financial and control systems were not good enough to give them the margins they needed.' Yet again: 'They have put too much emphasis on selling and not enough on product development.'

Old hands in the Roneo organisation would probably accept that there was some validity in all these views, though they could also point out that the trading environment, particularly in furniture and business forms, had been unfriendly for much of the time. Whenever the economy went flat office furniture was likely to be the first to suffer and the last to recover. Moreover, during a period of rampant inflation they had been subjected to statutory price controls, and these had hit them hard, particularly in the operations of Compagnie du Roneo in France. Nor had luck run their way in other respects. Following earlier dislocations caused by floods at the Dartford Works, there had been two serious fires at the Romford factory and another at the Scissett factory of Hirst Buckley.

All these were factors hostile to good profit performances, but within Roneo there was also a feeling that the overall situation had been mishandled by Vickers. Put colloquially, the argument would run: 'When Vickers took us over we were pleased because we felt our survival was in doubt. We also thought that Vickers, as engineers, would improve our manufacturing expertise and facilities. In fact, they seemed to think they had taken over a furniture business. They didn't want to know about the office machines. They thought it would be enough to put in a few Vickers men and then leave us to it. Worst of all, they chopped and changed top management so that we neither had consistent leadership nor consistent policies.'

Again, these were views carrying a degree of validity. When the office equipment group was formed the man put in charge was Robert Wonfor, whose equable personality and long experience as marketing and sales director of the Engineering Group ensured that the pains of assimilation would be minimised, the organisation kept in working order and the Vickers policy of growth pursued, as exemplified by the acquisitions of Compagnie du Roneo and Hirst Buckley. He was,

14

however, approaching retirement age and the question of the succession was vitally important. In January 1972 Vickers announced that 'in view of the considerable degree of common interest existing in the operations of the Roneo Vickers Office Equipment Group and the Vickers Limited Printing Machinery and Supplies Group it has been decided to integrate the administration and organisation of the two groups to the extent necessary to avoid duplication of effort and to achieve the maximum amount of co-operation and efficiency in sales, research and manufacture.' This meant that Colonel Jelf, then the Printing Machinery Group chairman, would also become chairman of Roneo Vickers in June 1972 though Robert Wonfor had agreed to defer his retirement for a year from that date in order to 'advise and assist' in carrying forward the integration policy. Sound arguments undoubtedly existed for such a policy since a good deal of affinity existed between the two groups, particularly between Roneo Vickers and Howson-Algraphy, representing the supplies arm of the printing machinery group. In the event, the policy was to be implemented only briefly and in part. An announcement in May 1972 said that a new managing director had been appointed with effect from early June. This was Derek Broome, who had held a series of senior appointments with GEC. His tenure of office was to last less than two months. Towards the end of August he left the Company as the result of clashes in approach and attitudes. L. G. Gooch, OBE, the production director, was called upon to fill the gap, with the title of deputy chairman. 'Len' Gooch had begun his Vickers career with Supermarine and was renowned for the fierce determination and dedication he brought to all the jobs he tackled. As managing director of Dartford Works he had a good knowledge of the steel furniture business and of the Roneo organisation, on the furniture side, even before the Vickers acquisition in 1966. It had therefore made sense for him to move from the Engineering Group to Office Equipment when the acquisition took place. A further change in the situation took place early in January 1973 when it was announced that Colonel Jelf was relinquishing all his executive appointments as a prelude to retiring from the Board in June on reaching his 60th birthday. With Robert Wonfor also nearing the end of his extra year as chairman, yet more changes had to take place. These brought in as chairman and chief executive D. J. Groom, Vickers young director of finance, an appointment which was well received in the upper reaches of Roneo management and seemed to mark the end of a period of considerable uncertainty and unhappiness. With him as commercial director came A. B. Butler, who had been one of his principal aides in the reorganisation of the Finance Department at

Vickers House and had subsequently gone to Vickers Medical Engineering as managing director. The Groom-Butler partnership clearly signified that the financial and commercial organisation within Roneo Vickers would be overhauled with the same thoroughness as in Vickers itself. With this change, L. G. Gooch took over particular responsibility for the Roneo Vickers activities in automated filing systems, notably the Halifax installation, and subsequently moved with this business when it was transferred to Engineering Group in April 1974: unhappily he died suddenly in 1976 during the year before his retirement.

Hopes of a new era of stability in Roneo Vickers top management were frustrated after little more than a year when, in June 1974, D. J. Groom was offered an important appointment in Canada and left the Company. Now, for the first time, a successor was found from within the ranks of the old Roneo company rather than by importation. The choice fell on K. W. Ketteringham, who at the age of 46 had 20 years with Roneo behind him. He had originally joined Roneo Neopost, the mailroom business, as a service mechanic, moved to the sales side and became Neopost's general sales manager in 1967 and then in 1972 marketing director of the Office Equipment Group as a whole. Given his strong personality and proved competence this was a very good appointment since it rallied behind him all those who felt that at last the business was to be led by someone who knew it intimately and whose thinking about its future was based on experience in the industry.

At the centre of this thinking was the need for a complete restructuring of the organisation. Hitherto Roneo Vickers had been seen as a unitary business, run from headquarters at Croydon, with one-half engaged in manufacturing and the other in selling, and with 38 executives reporting directly to the chief executive. The need, as Ketteringham saw it, was to recognise that Roneo Vickers comprised a number of highly specialised businesses which should be organised divisionally with each division made an effective sales and profit centre. This thinking he proceeded to implement so that in 1975 the organisation was re-shaped into four divisions—Roneo Vickers UK (managing director: L. M. Wilkey), Business Forms (managing director: H. Micklethwaite), Roneo Vickers France (director general: J. Dupuis) and Overseas Division (general manager: S. C. Dovey). The Group Board now had only five members in addition to the chief executive. They were Wilkey, Micklethwaite, and Dupuis, plus two functional directors, A. B. Butler, continuing to play a key role as commercial director, and P. J. Macleod as technical director. Wilkey and Macleod, like Ketteringham, had been with Roneo Limited before its take-over by Vickers. Micklethwaite came into the Roneo Vickers orbit as managing director of Hirst Buckley when it was acquired in 1969. Similarly, Jean Dupuis came to the centre of the stage in 1969 with the acquisition of Compagnie du Roneo. Tony Butler, as already noted, moved to Croydon in 1973.

Statistics of the Roneo Vickers empire in 1975 showed that, as measured by turnover, Roneo Vickers UK and Roneo Vickers France were much the most important of the new divisions, in the order of £30m to £40m, with Overseas Division in the order of £15m and Business Forms £7.5m. Overseas sales (exports from the UK plus sales by Compagnie du Roneo and other overseas manufacturing centres) in relation to sales to UK customers were in the proportion of 60 per cent to 40 per cent. In terms of product turnover, furniture and filing systems (plus partitions) represented just over one-half, office reprographics and supplies one-third, business forms 8 per cent and mailroom equipment 7 per cent.

These figures included results from six acquisitions made in 1974 and 1975, three to add to product range and output and three to strengthen the overseas sales organisation. The three manufacturing companies were Fanfold (business forms) in England, Behin Robustacier Meubles (office furniture) in France, and Mathias Bauerle (table-top offset duplicators and mailroom equipment) in Federal Germany. The three sales organisations were Roneo Vickers (Canada), Berg Bolinder (Sweden) and the Roneo division of Facit-Addo Inc (USA).

Most costly of the six was Mathias Bauerle at just under £3m. Its significance lay in the entry it gave into office printing machines as distinct from copiers and duplicators. Roneo had been founded on duplicators, and these remained a major product, but to compete successfully in the office machinery business it was necessary to be able to offer also both copiers and small printing machines. To start from scratch in developing its own products would have been impossibly costly as well as time consuming, and this meant either acquisition of established manufacturers or licence agreements. In the instance of copiers the chosen path was by licences, notably for the range of Saxon electrostatic photocopiers, but this, of course, meant that effectively two profits had to be earned.

The cost of the Fanfold acquisition was £750,000. With the addition of Fanfold products to those of Hirst Buckley it could be claimed that Roneo Vickers was now one of the largest British manufacturers of business forms, other than computer stationery. A new factory for Hirst Buckley was also completed in 1975 at Park Mill, near Scissett, at a cost of £1m. The project had been authorised at the end of 1973, not only to replace capacity lost in a fire at the existing works, but also 'to meet projected market growth until at least the end of 1978'. Unhappily, so far from the projected growth taking place, the business forms industry suffered more severely than most from the recession caused by inflation. A number of firms in the industry went out of business, and Roneo Vickers had to reduce and rationalise its own capacity, closing the Fanfold factory at Edmonton so that production could be concentrated at Park Mill. Even so the business made a loss in 1976. It was nevertheless in a position to take rapid advantage of the improvement in demand which was bound to take place and fortunately the omens in 1977 were more propitious.

Of the acquisitions to improve overseas sales outlets, that of Berg Bolinder, a major office equipment marketing company in Sweden, was made at a cost of some £500,000. Its results in 1975 and 1976 were rated as highly encouraging. The division of Facit-Addo Inc, which already marketed Roneo Vickers business machines in the United States, was bought for $1.9m. It now became the nucleus of a new company, Roneo Vickers Incorporated, which in 1976, its first full year of trading, was reported to have produced 'excellent results in the form of profits and increased sales of UK manufactured machines'.

Altogether, in the first ten years of its existence, the resources of the Office Equipment Group had been enhanced by 17 acquisitions — in 1968 those of Hadewe BV of Drachten, Holland, manufacturers of folding machines, and four companies in Australia which had been Roneo agents; in 1969 of Compagnie du Roneo and Hirst Buckley; in 1971, a French company manufacturing power-operated rotary filing systems and a Belgium company which had been Roneo agents in Belgium since 1907; in 1972, a Dutch company previously agents for Roneo reprographic equipment and for office furniture from Compagnie du Roneo, and Ertma SA, a Swiss company manufacturing mailroom inserters; in 1974, Fanfold, Berg Bolinder, and the group's former distributor in Canada; and in 1975 Behin Robustacier Meubles in France, Mathias Bauerle in Federal Germany, and Facit-Addo in the United States. The total cost of these acquisitions was in the order of £12.5m. In addition, a good deal of internal investment was undertaken, notably the new factory for Hirst Buckley at a cost of £1m and the project for complete rebuilding and re-equipment of the Romford factory at a cost estimated in 1975 at £4.5m.

To some extent the effort and resources put into reorganisation in the years 1974 to 1976 lay behind the relatively poor profit performance.

In a magazine interview, Ketteringham was quoted as saying: 'You can't have a marvellous bottom line while you're doing this. But I know that when I have gone through this painful transition period I shall have a stronger and better company at the end of it.' For more private consumption he might have added that the profits line was also diminished by Vickers Limited Head Office charges, with the prospect that the Roneo Vickers share would increase following the nationalisation of the Shipbuilding company. There could be no doubt, however, that with a turnover of over £100m, and more than half of this in overseas sales, Roneo Vickers had by 1977 fully achieved the objective declared by Sir Leslie Rowan in 1967: 'We intend to develop the Group to become an international organisation.'

5

Down to the Sea Bed

It would be difficult to say in whose eye the first twinkle appeared which led to the birth of Vickers Oceanics, but certainly it was Sir Leonard Redshaw who guided the pregnancy and made sure that a healthy child was born and grew up healthily.

The use of submersibles as work boats on the sea bed, as distinct from use in marine research, developed in North America. In the United States Perry Oceanographics of Florida designed and manufactured work boats in the late 1950s, chiefly to carry out survey work for the offshore oil companies. Later a small Canadian firm, International Hydrodynamics, of Vancouver, also entered the business with a series of boats christened *Pisces*. Barrow interest in setting up submersible services in Britain arose essentially from its knowledge and experience in building submarines and study of their uses and possibilities. This interest was quickened with the discovery of oil and gas resources in the North Sea, and also perhaps because a new outlet was being sought for the skills and expertise accumulated at Barrow. Initially, the customers were seen mainly as the international cable companies, the Ministry of Defence, and research institutions, though the likelihood of a developing demand by the oil companies provided a strong incentive. For the cable companies the submersible could offer an invaluable service in burying telephone cable on the sea bed, out of reach of trawling gear, and also in repair work. For the Ministry of Defence (Navy) it was the most efficient means of recovering torpedoes used in firing practice.

With these immediate purposes in mind the go-ahead was given in 1968 for an arrangement with International Hydrodynamics, to whom financial support was given, for use of a *Pisces* two-man submersible, and for the purchase of a ship from which to operate the submersible. For lifting *Pisces* in and out of the water Vickers designed special gear for building into the support ship, a stern trawler. Named *Vickers Venturer*, the support ship was operated by James Fisher & Sons Limited, a firm of ship-owners based at Barrow, with whom Vickers had a close and friendly relationship, though the submersible itself was manned by Vickers personnel. Early charters showed that the project was well conceived, and contracts for cable laying were successfully carried out in the Atlantic and in the Messina Straits in the Mediterranean, as also were contracts for the Navy in torpedo recovery. When a Viscount airliner crashed in the Irish Sea, *Pisces* was also called in to locate the wreckage.

The first mention of these developments in Vickers annual report came in 1970 in a single sentence. It said: 'Other activities included developments in the oceanics field, where the *Pisces* submersible has been employed on work for the Ministry of Defence, fisheries, geological institute and the National Environmental Research Council.' A one-sentence reference appeared also in the 1971 report: 'In oceanics, the *Pisces* submersible was in increasing demand for underwater search and research and a second submersible was accordingly put into commission.'

Enough had been achieved, in fact, to set up a full-scale commercial operation. Not least it had been shown that the *Pisces* could be kept in operation in rough conditions in the North Sea when normal diving operations were impossible. In March 1972 it was announced accordingly that a new company, Vickers Oceanics Limited, was being formed, with Vickers holding 63 per cent, the National Research Development Corporation 26½ per cent and James Fisher & Sons 10½ per cent. The company would initially operate three *Pisces* submersibles from two support ships, the *Vickers Venturer* (600 tons gross), already in service, and the *Vickers Voyager* (2,850 tons gross), a converted factory trawler bought in Norway. By the end of 1972 this fleet was in operation, and contracts included inspection of pipelines in the North Sea.

Though all was going well Vickers knew that the business had still to prove its commercial viability and that some would regard it as a risky venture. In an article in the *Financial Times*, C. G. Milner, marketing development manager of the Shipbuilding Group, wrote: 'It is not a business for the weakly capitalised or inexperienced enthusiast. Capital and operating costs are high. The whole business is highly geared and the risks of financial failure are significant.'

The moment of truth came dramatically in August 1973. *Pisces III* and her two-man crew, engaged in a cable-burying operation off the southern Irish coast, became stranded on the ocean bed at 1,575 feet. No rescue had ever been made at a depth as great as this. For 2½ days television, radio and press reported hour by hour on the rescue operation. The only factor that mattered at that stage was to save the two

Helicopter view of Millbank Tower nearing completion.

Three Vickers Chairmen at a ceremony in 1963 to mark the opening of Millbank Tower—Lord Knollys (1956–62) on left, Sir Leslie Rowan (1967–71) and Sir Charles Dunphie (1962–67). Lady Knollys is unveiling the plaque and Lady Dunphie is also in the group. Almost hidden behind Sir Charles is Lady Rowan.

First meeting in Millbank Tower of the Board of Vickers Limited. From l. to r.: Dr. W. D. Pugh, Sir Sam Brown, G. H. Houlden, Colonel A. T. Maxwell, A. O. Bluth, D. L. Pollock, A. H. Hird, R. P. H. Yapp, W. D. Opher, Lord Knollys, Lord Bicester, Sir Leslie Rowan (Managing Director), Sir Charles Dunphie (Chairman), A. M. Simmers (Secretary).

Sir Charles Dunphie in attendance on Her Majesty The Queen Mother who launched the liner "Northern Star" at Naval Yard, Newcastle, in 1961, despite a sprained ankle.

Lord Weeks (Chairman 1949–56), on left, with Sir Keith Smith, navigator of the Vickers Vimy which made the first England–Australia flight and later Vickers' representative in Australia.

Sir Peter Matthews, chief executive of Vickers from 1970, opens a Vickers computer centre at Newcastle.

Jim Hendin, Chairman of Vickers Engineering Group from 1970, and also Assistant Managing Director, Vickers Limited, from 1976.

Eric Harrington, chief executive of Canadian Vickers from 1967, and Peter Scott Maxell, chief executive of Vickers Australia also from 1967, meet at Vickers House.

men, Roger Mallinson and Roger Chapman, but behind the human drama almost certainly lay the future, not only of Vickers Oceanics, but of submersible operations generally in British waters. If the rescue attempt failed there would be at the very least a severe setback to the development of submersibles as work boats. If it succeeded, though the immediate cost would be high, the longer-term outcome would probably be firmer confidence in submersible operations and with it enhanced business for Vickers Oceanics. For the situation to be so finely poised probably owed more to emotion than to logic, bearing in mind that the lives of individual divers were frequently being lost in accidents in the North Sea, but its reality had to be recognised.

The accident occurred while *Pisces III*, operating from *Vickers Voyager*, was engaged in burying a section of transatlantic cable 150 miles south-west of Cork. On the morning of Wednesday, 29th August, Roger Chapman and Roger Mallinson had surfaced shortly after 9 o'clock at the end of a stint of work. Both were experienced submersible pilots. Before joining Vickers Oceanics Chapman had been, among other things, the navigator of a nuclear submarine. Mallinson, formally a computer technician, joined Oceanics in 1971, first undertaking engineering work and then became a pilot.

As *Pisces* surfaced, a *Gemini* inflatable boat came from *Voyager* and attached a tow line so that she could be hauled in and lifted aboard. All appeared to be proceeding normally when suddenly Chapman and Mallinson became aware that *Pisces* was dipping sharply backwards. She was in fact sinking rapidly by the stern. At 175 feet below the surface the tow rope held for a few minutes but then snapped, and *Pisces* continued her plunge stern-first to the sea bed. At 1,575 feet on the depth gauge she finally hit the bottom and became embedded in the mud by the stern.

By good fortune neither man had been injured and all systems, including radio-telephone communication with *Voyager*, remained intact. For Chapman and Mallinson it was therefore a matter of re-arranging equipment and fittings inside the submersible to adjust to the 90 degrees stern-down attitude, of minimising physical movement and talking in order to conserve oxygen, and of having confidence that their colleagues on the surface would be making an immediate all-out effort to rescue them.

Quick calculations showed that oxygen supplies should last until 1600 hours on Saturday so that there was a period of a little less than 3½ days in which to complete the rescue operation. The rescuers had first to find *Pisces III* on the sea bed and then get a line or lines attached to her strong enough to haul her to the surface. 'Strong enough' meant very strong indeed, preferably two lines rather than one, since *Pisces* had to be pulled out of the mud and was heavily down at the stern. The accident had been caused by flooding of a rear compartment when the tow line from *Voyager* fouled the hatch locking mechanism and tore off the hatch. This open hatch was, in fact, to prove important in the rescue operation since a 'toggle'—a form of grapnel opening out like an umbrella—could be inserted into it at the end of a rescue line.

In his book *No Time on Our Side*, Roger Chapman wrote: 'We never guessed what huge preparations and dynamic efforts were already being made just one hour after the accident.' By then Sir Leonard Redshaw and Commander Peter Messervy, general manager of Vickers Oceanics, had already met and set in train the first rescue moves. Redshaw had vivid memories of the *Thetis* tragedy in Liverpool Bay in 1939, when three Vickers men were among nearly 100 who lost their lives. *Thetis*, built by Cammell Laird, was engaged in diving trials, with a naval complement of 53 and 50 'observers', most of them civilian technicians. The inquiry showed a chain of human errors, compounded by slowness in communication and a failure to centralise the rescue operation quickly enough in knowledgeable and experienced hands. Remembering *Thetis*, Redshaw knew that action must be taken instantly to set up a control centre and to mobilise every possible rescue facility, desirably in duplicate. With a Base Operations Room set up in Barrow under the direction of Redshaw and Greg Mott, managing director of Vickers Oceanics, who had rushed from London to Barrow, Peter Messervy flew to Cork to take charge at the site. There could have been no man better qualified for the job. Before joining Oceanics Messervy had been a submarine officer in the Royal Navy: moreover, his experience with *Pisces* submersibles included an incident in Vancouver Bay when he himself was in a submersible rescued from 600 feet.

The detailed log of the operation shows that within two hours of the sinking an instruction was sent to *Vickers Venturer*, operating in the North Sea with *Pisces II*, to return to nearest port in order that *P.II* could be airlifted to Cork, a request sent to International Hydrodynamics Limited in Vancouver to fly another *Pisces* submersible to Cork, and the emergency services of the Ministry of Defence alerted. At noon the US Navy also offered help, in particular by making available the unmanned submersible CURV (controlled underwater recovery vehicle). Thus within a few hours of the accident arrangements had been made for two *Pisces* submersibles and CURV to be flown to Cork, and for surface vessels to sail to the location of the sinking so that one of them could take over on-the-spot duties, especially in communication, while *Voyager* returned to Cork to pick up the rescue submersibles. The first ship to arrive was the Royal Fleet Auxiliary *Sir Tristram*, and having transferred the communications team *Voyager* set off for Cork.

Speed in assembling the rescue submersibles was vital, and in the North Sea *Pisces II* was crane-lifted from *Vickers Venturer* to *Comet*, an oil rig supply ship which could steam four knots faster. With her went four of the operations team. *Comet* headed at full speed for Teesport and *Pisces II* was then road-freighted to Middlesbrough Airport where an aircraft was waiting. Meanwhile, at Halifax in Nova Scotia, *Pisces V* was taken off cable-burying operations to be loaded on board a Hercules of the Royal Canadian Air Force for the trans-atlantic flight. The flights from Middlesbrough and Halifax arrived at Cork during the early hours of Thursday, within an hour of each other, so that *Pisces V* and *Pisces II* were ready for shipping when *Voyager* docked at 8.15 a.m. Two hours later *Voyager* set sail. CURV was also on its way by air to Cork and to pick it up there was a Canadian cable ship, the *John Cabot*.

The role of support ship had now been taken over from *Sir Tristram* by HMS *Hecate*, a small survey ship, which remained on site and gave invaluable help even after *Voyager* had arrived towards Thursday midnight to resume control of the rescue operation. Unhappily, the weather had now become dangerously rough and under normal circumstances there would have been no question of launching *Pisces II* in these sea conditions. Nevertheless, preparations for launching went ahead, only for another problem to appear. Three outboard engines, providing power for the *Gemini* inflatable boat, all refused to start, and it was 0230 before *Pisces II* could begin her descent, carrying a line with toggle attached. Two toggles—in effect folding grapnels—had been hurriedly made at Barrow from a special design by Vickers Oceanics and flown to Cork. The first task was to locate *Pisces III* by use of sonar equipment, an operation requiring systematic and patient searching. Misfortunes were accumulating, however, and the search was still under way when *Pisces II* developed a leak. Though at that stage it was not a major leak a return to the surface could not be avoided, a critical delay with not more than 36 hours remaining before oxygen supplies on *Pisces III* ran out.

Pisces V from Canada had now to take over the search and rescue role. She was launched at 0545 and reached the sea bed 45 minutes later. For six hours she searched without success. 'It was almost un-believable that she was unable to locate us,' wrote Roger Chapman.

'No one knew that by a one in a thousand chance we had landed stern first in a slight hollow, where we were hidden from the searching sonar, except at short range.' Again the search had to be abandoned.

Where high hopes had been held for a rapid rescue, the situation was now becoming tense. Captain Edwards of the *Voyager* believed, however, that he could station his ship almost directly above the lost *Pisces* and from this position *Pisces V* was launched a second time. This time, at a distance of 800 feet, her sonar locked on to *Pisces III* and at 12.44 p.m. she came alongside the upended submersible, her lights and proximity putting new heart into Chapman and Mallinson who had now been incarcerated for 51 hours. Fixing her snap hook on *Pisces III* proved to be a difficult operation, and one failure to get a firm hold was followed by attachment of the hook to the propeller guard, too flimsy to be an acceptable lift point. For over an hour *Pisces V* tried unsuccessfully to move it to the correct lift point, and then, with her batteries running low, she was instructed to remain alongside while on *Voyager* a choker was devised to run down the line with a second line attached.

Another attempt to launch *Pisces II* in very rough seas showed that she was still unserviceable, but *John Cabot* with CURV on board had now arrived on the scene. *Pisces V*, after her long vigil at the side of *Pisces III*, was forced to resurface: she and her Canadian crew had been under water for 18 hours.

Work had meanwhile been proceeding on repair of *Pisces II* and this time her launch was successful. Equipped with a toggle at the end of a 3½-inch polyprophylene line, she arrived beside *Pisces III* at 0500 on Saturday and quickly fixed the toggle in the open hatch. There was thus two lines attached, though there could be no confidence in the line fixed to the propeller guard. It was now a question of whether to begin the upward haul immediately, using the line fixed by *Pisces II*, or to use CURV in an attempt to make it a belt and braces operation.

Some repair work had proved necessary on CURV so that her earlier participation had not been possible, but it was now reported from *John Cabot* that she was ready for launch. Equipped with the second toggle and a 6-inch braided nylon line, CURV began her descent at 0940, following the line left by *Pisces II*. Fifty-five minutes later she had fixed the toggle in the open hatch.

In the stranded submersible Chapman and Mallinson had to an extent become comatose, with severe headaches, and were barely aware of these latest developments. At 10.50, however, they acknowledged a call to tell them that the lifting operation was about to begin. 'Yet the fact that we were actually going to be lifted at last still did not register fully in my mind,' Chapman recalled. 'There was certainly no thought that rescue might be just round the corner, but instead there was annoyance that we were to be disturbed.'

The lift took 155 minutes, with *Pisces III* twisting and turning violently to the great discomfort of the two exhausted men. She surfaced just after 1 p.m. on Saturday, 1st September. On the latest estimates, her oxygen supplies would have given out five hours later. Altogether she had been below the surface for 84½ hours. Never before had a rescue been accomplished from such a depth, and the world applauded a remarkable feat of organisation, international teamwork and individual dedication. Many people had taken part, some in dangerous conditions and to the point of exhaustion. Roger Chapman dedicated his book to 'Peter Messervy, Bob Eastaugh and their magnificent rescue team from many parts of the world who battled without sleep for 3½ days to bring Roger Mallinson and myself to safety from the darkness of the Atlantic deep — back to our wives and families.' Of Chapman and Mallinson Lord Robens asked: 'How does one describe the quiet heroism and quality of men, sealed in their tiny craft for over 80 hours, finally rescued with the odds against them, who continue to go on diving again?'

Following the rescue Vickers Oceanics set up an Inquiry Panel which was joined by representatives of the Department of Trade and Industry, the American Bureau of Shipping and Flag Officer, Submarines. Its purpose was to establish the cause of the accident and to review the procedures carried out in the rescue. The cause was confirmed as fouling of the hatch locking mechanism leading to dislodgement of the hatch. 'A contributory factor', said the Panel's report, 'was the removal of a locking bar for the hatch mechanism by the operating personnel, six weeks prior to the accident, because of the difficulties they were then encountering with small leaks. This was not an approved change to the configuration of the submersible, but in many respects it was consistent with the configuration of *Pisces II*, which had operated successfully and without incident for a period of three years.' A Ministerial statement in the House of Commons attributed the sinking to a failure to adhere strictly to established procedures rather than to any inherent fault in the vessel itself, commented on the 'remarkably good' record of the *Pisces* type of submersible and noted that measures had been taken to provide greater security for the hatch locking mechanism.

Among points arising from the examination of the rescue procedures was a complaint by Captain Edwards of *Vickers Voyager* that the work of rescue had been handicapped by a trawler, carrying a BBC reporting team, which persisted in remaining dangerously close to the point of rescue: not only was his own ship put at hazard but also the marker buoy, while propeller noise interfered with sonar and underwater communications. After requests by VHF and signalling lamp had failed to persuade the trawler to withdraw to a safer distance a helicopter was flown from HMS *Hecate* carrying a blackboard with a chalked message, but this, too, was ignored. A further complaint against the BBC was the interception of a message from Roger Chapman saying that Roger Mallinson's condition had deteriorated slightly though he was bearing up well, and the use of this message in a news bulletin couched in alarmist terms. Not only was this a possible contravention of the Wireless Telegraphy Act, but it caused alarm and dismay at a critical point in the rescue operation. Moreover, it was strongly felt in Vickers that BBC reporting of the rescue became hostile to the Company following a complaint by Sir Leonard Redshaw at a press conference about interference with the rescue operation. Putting aside these particular complaints, Vickers felt that a clear need had been established for agreed procedures in press, television and radio reporting of underwater accidents to ensure a full and rapid flow of information without causing hazard to the work of rescue.

It remained to count the financial cost of the accident and to assess its likely impact on commercial operations. The cost of the rescue operation itself, plus the loss of revenue from the interruption to work, considerably reduced Shipbuilding Group profits in 1973 (£2m as compared with £2.7m in 1972). Commercially, however, the successful rescue certainly did nothing to destroy customer confidence and, if anything, enhanced it. With these reactions apparent, Vickers was content to continue expansion of the oceanics company and a third support ship, *Vickers Viking*, became operational in 1974, followed by two more, *Vickers Viscount* and *Vickers Vanguard*. Further developments in 1975 and 1976 included enlargement of the submersible fleet to ten, setting up a full-scale operational base at Leith, formation of a joint company with Fred Olsen & Company to develop activities in the Norwegian sector of the North Sea and establishment of a diver lock-out capability. The submersible fleet included three built by Vickers-Slingsby with hulls constructed of glass reinforced plastic (a fourth was under construction in 1977). Appropriately this series was given the identification letters LR in honour of Sir Leonard Redshaw, the man who made the whole thing possible by bringing Vickers Oceanics into existence, by foreseeing the possibilities of adapting glass reinforced plastic to submarine construction, and by cajoling Vickers Limited into acquisition of the Slingsby company. The third of these GRP submersibles was a highly sophisticated vessel, incorporating a steel lock-out facility so that divers could leave and re-enter on the sea bed, and having the capacity to maintain five people underwater for seven days.

In 1976 Vickers bought from the National Research Development Corporation its 26.31 per cent holding in Vickers Oceanics Limited at a cost of £1,625,000. From the NRDC point of view this had been in many respects a model operation—the contribution of funds to a pioneering business and profitable withdrawal when the business had become viable. In 1977 Vickers also acquired the holding of James Fisher & Sons so that Vickers Oceanics became a wholly owned subsidiary. This meant that Vickers Limited had by 1977 invested some £15m in the Oceanics company.

Whether this investment would bring an adequate return had still to be established. Though sales and employment by Vickers Oceanics had steadily increased, its return on capital had no less steadily diminished. Reasonable profitability had been achieved in 1974 and 1975, but in 1976 and 1977 a number of adverse factors appeared. In particular, competition intensified so that margins became low. 'The long term prospects look good,' said the 1976 annual report, but the short-term hiccup was uncomfortable. To pioneer and innovate inevitably carries commercial risks, but these are risks that within reason the great companies must be prepared to take if they are to preserve their own self-respect and credibility and also serve the national interest.

In addition to submersible services, Vickers had interests in other aspects of offshore engineering, as was recognised with the formation of an Offshore Engineering Group in 1975. Joining Vickers Oceanics in the new group were Vickers Offshore Developments, Brown Brothers and Vickers-Slingsby.

Offshore Developments, like Vickers Oceanics, had been nurtured by the Shipbuilding Group. It was the repository of accumulated skill and knowledge in the management of offshore operations, particularly where innovation was required. To strengthen the services it offered, Vickers in 1976 acquired Aylmer Offshore Limited, a company operating design engineering and contributory services in the offshore oilfields, and also set up Vickers-Intertek Limited to develop a system to encapsulate wellhead equipment on the sea bed. A further reflection of innovative thinking was a project to develop a method of explosive welding under water, first begun at Barrow and continued at the International Research & Development Company. In 1977 the enlarged organisation was formed into a new company, Vickers Offshore (Projects and Developments) Limited, with offices in London and Barrow.

Innovation was no less a feature of the activities of Vickers-Slingsby. In addition to its work on GRP submersibles Slingsby continued to design and build sailplanes, its original activity, and was also engaged in the manufacture of the Vikoma Seaskimmer, a device developed in conjunction with BP for dealing with surface pollution. Nor was the ability to innovate lacking at Brown Brothers who within a period of nine months in 1973 designed, manufactured, tested and delivered to Honolulu heave-compensation equipment required urgently by the Scripps Institute of Oceanography. The equipment was for installation in a drilling ship engaged in exploration of the ocean beds. The Scripps Project Leader said he had never known equipment as sophisticated as this to be so quickly designed, patented and manufactured. For Brown Brothers this was a valuable new product to add to the steering gear and stabilising equipment for which it already had an international reputation.

Though Sir Leonard Redshaw retired from executive duties in 1976, he continued to preside over the Offshore Engineering Group as non-executive chairman until the end of 1977. Emerging during 1978 as chief executive of the Group was Dr John Rooke who had come into the Vickers orbit with the acquisition of Brown Brothers. His experience in industrial management was backed by an unusual combination of academic achievements, a Doctorate of Philosophy having been added to a B.Sc. Other Offshore Engineering directors included G. E. Burton, managing director of Vickers-Slingsby and recognised as one of Britain's leading glider pilots.

6

Vickers Overseas

Strong performances by the Australian companies and by Canadian Vickers were a feature of the mid-1970s, all the more noteworthy because they were achieved in adverse conditions. Canadian Vickers in particular found itself trading in an increasingly difficult environment, not only because of inflation in the Canadian economy and militancy in industrial relations, but also because of political pressures created by the French-speaking separatist movement in Quebec province. Compared with the operations in Australia and Canada, Vickers had only a small presence in South Africa, but reorganisation of a considerable diversity of activities, coupled with expansion in some, began to make it more significant. In India, Vickers had its 24.9 per cent holding in ACC-Vickers-Babcock Limited, from which it gained little advantage, and with prospects unpromising in defence equipment and in trading activities withdrawal rather than expansion was the order of the day.

To assess the overall contribution of the Australian and Canadian companies, analysis of results for the four years 1973 to 1976 shows that together they accounted on an upward trend for between 16 and 20 per cent of sales of the whole Group and 16 and 25 per cent of profits. In absolute figures, sales increased from £36.8m in 1973 to £77.9m in 1976 and profit went from £2.2m in 1973 to £6.4m in 1975, followed by a small decrease to £6m in 1976. As between Australia and Canada, the bigger contribution in sales came from the Australian companies, but in profits over the four years Canada contributed £9m and Australia £8.8m. Both had a good level of return on capital employed, Australia in the order of 18 per cent in 1975 and 1976 and Canada a remarkable 46 per cent in 1975 followed by 27.8 per cent in 1976. These were appreciably better figures than for the UK Engineering Group, though in 1976 the latter had moved to 15.9 per cent from very low figures in the early 1970s. On the yardstick of numbers employed, Canada was in the order of 1,500 and Australia 3,900 within the Group total of just under 42,000 in 1976. With the nationalisation of shipbuilding in 1977, however, the Group total fell to some 28,000 so that employment in Australia and Canada together represented about 18.5 per cent. By any standard of measurement, therefore, the operations in Australia and Canada had a substantial role in Vickers affairs.

In Australia the 1970s saw considerable expansion and reorganisation, which against a background of lean years in the economy was a policy requiring shrewd judgment, based not only on figures and forecasts, but also on business instinct. In 1972, the omens seemed in no way propitious. Redundancies were necessary and Vickers Hoskins in Western Australia suffered a loss. Nevertheless in 1974 activities in Western Australia were expanded by the formation of two new companies. At Kalgoorlie Vickers Hoskins (80 per cent) and Keogh Brothers (20 per cent) set up Vickers Keogh Pty Limited to undertake major overhauls of heavy mining equipment for gold and nickel mining companies in the Kalgoorlie region. This was a development based on the existing business operated by the three Keogh brothers. This was not, incidentally, the first time that the Vickers and Keogh names had been linked. Two generations earlier, in England, the Keoghs' grandmother had been a member of Albert Vickers household staff. At Bassendean, on the outskirts of Perth, where Vickers Hoskins and Vickers Hadwa were already located, a new company was established by Vickers Hoskins and Mt Newman Mining, on a 50–50 basis, under the title W. A. Mining Engineering Services Pty Limited. Its purpose was to develop facilities for the repair and maintenance of heavy diesel engines and locomotives for the mining industry. The two new companies quickly proved successful.

Also in 1974 an acquisition of great significance was made in eastern Australia – part of the assets of the Hadfield-Goodwin-Scotts Group which had gone into liquidation. Its significance lay particularly in the entry thus provided into the railway transportation business and the manufacture of coal processing plant. The new assets were integrated with those of Vickers Ruwolt and the 1974 annual report recorded immediate benefits – 'large orders for cast steel bogies at Richmond, draft gear at Sydney steel foundry and railway freight wagons at Scotts of Ipswich in Queensland'. Vickers Hadwa also received major orders for cast steel bogies.

The cost of these developments in 1974 was relatively small, the total being less than £2m. Other new enterprises were also undertaken. Vickers Ruwolt formed a subsidiary, Vickers Research Pty Limited, to tackle design and development problems in the industries served by the Vickers companies. The possibilities were explored of forming manufacturing and service companies outside Australia. This led to the setting up in Singapore of a foundry designed to produce both iron and steel castings: the project received support from the Singapore Economic Development Board before becoming a company under the title Uniteers Vickers Pty Limited, jointly owned by Vickers Australia and United Engineers. Twelve miles from Singapore, on the Indonesian island of Batam, Vickers Australia also went into business with Pertamina, the Indonesian State oil enterprise, in forming a company, P. T. Patra Vickers, to provide engineering and maintenance services in the area, particularly to the oil and gas industries. Also looking northwards, Vickers Hoskins were pursuing in 1977 a project to set up in Malaya a joint company to manufacture boiler plant to burn palm refuse in the palm oil industry, a product which Hoskins had developed over the years with notable success. Across the Indian Ocean to the west lay South Africa, where the mining industry had long provided an export market for Ruwolt products. In 1977 agreement was reached in principle for Vickers Australia to set up a specialised steel foundry in the Bantu Homelands of South Africa. Its purpose was to employ there the advanced techniques developed in Australia in the manufacture of steel castings, particularly those incorporated in Symons crushers, a product licensed from Nordberg of Milwaukee. Formation of a company with Lennings Holdings Limited as partners, with a 45 per cent holding, was encouraged by financial support and incentives offered by the Bantu Investment Corporation and the South African Government.

Another facet of growth in Vickers Australia was extensive internal investment to add to productivity by introducing the most modern machine tools. The replacement programme was accelerated in order to take full advantage of the Government incentives on depreciation and investment, and in April 1977 the managing director, Peter Scott Maxwell, reported that orders to the value of £2.7m had been placed for new machine tools.

He was also able to report that, despite the economic recession in Australia, 1976 had seen an advance in profits from £2.58m in 1975 to £2.97m. This, he said, was 'largely due to the Group's very wide product range and geographical coverage within Australia, materially assisted by substantial export contracts undertaken by Vickers Hoskins to New Zealand, Malaya and Indonesia, and by Vickers Ruwolt to South Africa'. Particularly important to Vickers Hoskins at this time were contracts from New Zealand for Portainer cranes: with these and earlier orders, cargo handling cranes manufactured by Vickers Hoskins could be seen in operation in all the main ports of Australia and New Zealand.

Organisational change during these years saw Vickers Ruwolt, Vickers Hoskins and Vickers Hadwa become operating divisions of Vickers Australia Limited with effect from the beginning of 1975. Vickers Australia was itself under the overall policy direction of Vickers Holdings Pty Limited, with headquarters at Melbourne, as were Vickers Cockatoo Dockyard Pty Limited and Vickers Sales Pty Limited. This last company was formed at the beginning of 1976 to provide general oversight of the activities of the companies already existing to sell in Australia the products of Crabtree-Vickers, Howson-Algraphy and Roneo Vickers. The activities of the Crabtree-Vickers sales company had earlier been reinforced by the acquisition for £380,000 of Middows Brothers, a company engaged in supplies and services to the printing and graphic arts trades.

Managing director and chief executive of both Vickers Holdings and Vickers Australia was Peter Scott Maxwell, who sat also on the boards of Cockatoo and Vickers Sales and was thus the central figure in Vickers activities in Australia. His was a considerable achievement. He had made himself totally acceptable to his Australian colleagues, aided no doubt by being a Scot rather than a Pom, and also perhaps because he so obviously enjoyed the Australian scene. From this base he had been able to guide Vickers activities in Australia to growing success despite the surrounding economic gloom. This success owed much also to the non-executive chairman, E. P. M. Harty, and the non-executive directors, representing strongly a wide range of authority and experience in Australian financial and industrial affairs. 'Ted' Harty, as much at home in England as in Australia, became chairman of Vickers Holdings in 1965. He was then managing director of Australian United Corporation, who had been Vickers financial advisers in Australia for many years. When subsequently he gave up executive duties with AUC, though remaining a non-executive director, he received many invitations to join the Boards of other companies and those he accepted included the Australian offshoots of British companies, not least Guest Keen & Nettlefolds and Legal & General Assurance Society. His non-executive colleagues on the Vickers Holdings board included three whose association with Vickers activities in Australia went back for many years – G. M. Bunning, a notable figure in the Australian timber industry, Sir Gregory Kater, director of a wide spread of Australian companies ranging from electrical engineering to banking, and J. G. Wilson, CBE, managing director of Australian Paper Manufacturers Limited.

At the operating levels in Vickers Australia the chief executives in 1977 were N. H. Ineson (Vickers Ruwolt), W. T. Peart (Vickers

Hoskins) and W. W. G. Meecham (Vickers Hadwa), all Australians with long service in these businesses. At Cockatoo the chief executive was customarily recruited from the Royal Australian Navy and when Captain Roger Parker, OBE, a notable servant of the Company, retired in 1971 he was succeeded by another naval captain, R. R. W. Humbley, who came to Cockatoo from the post of director-general, Naval Production, with the Commonwealth Government. As with Vickers in · Britain, long service was commonplace in the Australian engineering companies, sometimes spanning generations, as with Noel Ineson, whose father had been chief executive of Vickers Hoskins for many years. Sometimes it also spanned Britain and Australia, as exemplified by Scott Maxwell, but also at shop floor level as Lord Robens discovered during a visit to Vickers Hoskins in 1976. While touring the works he met a fitter and turner, Bill Shotton, who had worked with Vickers in Newcastle from 1935 to 1964 (apart from war service with the Royal Navy) before taking his family to Western Australia. Looking for a job there it seemed inevitable that he should gravitate to Vickers Hoskins, but what perhaps he had not expected was to find himself operating, among other machines, the identical borer he had operated at Scotswood. It had made the migration a few years ahead of him.

Possibly a 'family' feeling was inherent in Ruwolt and Hoskins even before they became Vickers companies in the 1950s, since both had begun as family businesses, one founded in 1902 by Charles Ruwolt to make windmills for the farms surrounding the country town of Wangaratta, in Victoria, and the other in 1895 as a blacksmith's business in Perth. Now it could be claimed that Ruwolts was one of the most important engineering establishments in the southern hemisphere, and Hoskins the largest heavy engineering company in Western Australia.

Cockatoo had distinctions of a different kind. In its setting, a 44-acre island in Sydney Harbour, no Vickers establishment was more picturesque. It also had a special place in Australian history. Originally the home of cockatoos, it was used by the early settlers at Sydney as a granary, with underground silos cut into the sandstone, and then as a penal settlement which developed by stages into a naval dockyard. A dry dock was first built on the island, using convict labour, in the 1850s, and named the Fitzroy Dock. For a time in the 1870s Cockatoo housed a reformatory for women, but with the building of a second dock, the Sutherland Dock, it blossomed into a fully operating dockyard run by the New South Wales State Government. In 1911 it was taken over by the Commonwealth Government which operated it until 1933 when a group of Sydney businessmen formed a company to take over management of the dockyard. Among them was Sir Keith Smith,* Vickers' representative in Australia, and it was probably on his advice that Vickers acquired control of the company in 1947, with Sir Keith as chairman and Engineer Captain G. I. D. Hutcheson as managing director. During the Second World War Cockatoo had performed outstanding work in repairing US warships damaged in the Pacific and in fitting out troop carriers, including the great liners *Queen Mary*, *Aquitania*, *Mauretania* and *Queen Elizabeth*.

Under Vickers control after the war its activities continued to be directed mainly to building and refitting ships for the Royal Australian Navy, though some commercial shipbuilding was undertaken, and a large heavy engineering facility was maintained. In the 1970s naval shipbuilding virtually ceased though repairs and refits, notably of Oberon class submarines, continued as an important workload, to the extent indeed that a new 'slave' dock was brought into operation in 1974 to enable two submarines to be refitted simultaneously. Engineering work included large and important contracts, for example a 800-litre bucket wheel dredger, launched in 1976 for the Melbourne Harbour Trust. Though in the nature of the Lease and Trading Agreement with the Commonwealth Government, Vickers had to expect more prestige than profit from managing Cockatoo, they accepted a new 21-year Agreement in 1972. In effect this made the Company an independent contractor, paying rent to the Commonwealth Government, though with clauses regulating naval work.

On a broad view, there could be little doubt that in 1977 Vickers had good reason to be pleased with the performance and prospects of the Australian activities, and with the harmonious and effective working relationships established across 12,000 miles. These were aided not least by visits to Australia by Lord Robens, Sir Peter Matthews (a member of the Vickers Holdings Board) and other senior executives from Vickers House (in continuation of well-established practice) and by return visits by the top Australian executives, of whom Peter Scott Maxwell was now a Director of Vickers Limited.

There could also be satisfaction with the current and recent performance of Canadian Vickers. Whether this could be sustained, however, given the increasing political turmoil created by the French-speaking separatist movement, was another matter. Much depended on whether discriminatory action would be taken, overtly or covertly, against English-controlled companies in Quebec province. In logic nothing would be gained in any quarter by diminishing the perfor-

*Navigator of the Vimy, captained by his brother, Ross Smith, which in 1919 made the first Britain–Australia flight.

mance of these or any other companies, but in some respects the omens were not propitious. It had to be hoped that in the long run good sense would prevail and, in the meantime, Canadian Vickers must pursue its own affairs while taking due regard to the political and emotional environment in which it operated. Certainly it had always been the practice to ensure that membership of the Board included leaders from the French-speaking business community, and a distinguished French-speaking Canadian, Wilbrod Bherer, QC, was chairman from 1967 until his resignation in 1976 on reaching the age of retirement. By then he had been associated with Canadian Vickers for 26 years, initially as chairman of the George T. Davie business at Quebec.

President and chief executive from 1967 was J. E. Harrington, a leading figure in the business and community life of Montreal (at one time a City Councillor) who came to Vickers from the Anglin-Norcross Corporation (Montreal), of which he was president. The company's increasingly good performance reflected his forceful, forthright and extrovert leadership. On the retirement of Wilbrod Bherer he became chairman as well as chief executive. Earlier, in 1974, he had been appointed a Director of Vickers Limited.

With the cessation of shipbuilding activities in 1969, attention had been focused on heavy engineering and shiprepairing, and both activities stood in good shape in the mid-1970s. The work of the Industrial Division fell under two main heads—defence and nuclear. In defence, the US-Canadian agreement on shared production resulted in large contracts, chiefly for submarine hull sections and torpedo tubes. In nuclear construction, the Division had exceptional experience and exceptional facilities for the building of reactors and other major components for the electricity generating industry. Not least, because of its extensive river frontage, it was able to assemble these massive items of equipment (some weighing in the order of 1,000 tons) and ship them complete. These facilities it enhanced in 1975 by erection of a new building specially designed for the work of assembly. The cost was £1.5m and the effect was to triple previous capacity. In the 1975 annual report it could be claimed: 'Canadian Vickers now has the only nuclear assembly facility of this kind and magnitude in Canada'. In 1976 work was proceeding on five reactors for the Canadian market and one for the Argentine, and in that year a reactor was also ordered for South Korea. In the Marine Division, the workload depended to an important extent on the refit programme for destroyers of the Royal Canadian Navy. Four refits were completed at the beginning of 1976 but there was then a gap of almost a year before the docking of the first of a further four destroyers.

As compared with its interests in Australia and Canada, Vickers investment in South Africa was small, though Roneo Vickers, Howson-Algraphy, Vickers Instruments, the Engineering Group and Vickers Australia all sought to sell in the South African market. To this end, a number of sales companies had been established, together with a small manufacturing activity by Roneo Vickers near Johannesburg. All these activities operated under the umbrella of Vickers Southern Africa (Pty) Limited, formed at the beginning of 1973 with headquarters in Johannesburg. Its non-executive chairman was initially H. H. McGregor, a well-known South African businessman, but on his death in 1974 the appointment was taken up by O. J. Breakspear, then chairman of Pilkington Brothers South Africa (Pty) Limited. His career with Pilkingtons had included senior appointments in the U.K. and Australia as well as South Africa, so that he had an exceptional range of experience.

During 1976 several projects for manufacturing activity came under examination, and this led to four small but significant developments almost simultaneously. Towards the end of 1976 Michell Bearings set up a small production facility in Boksburg. In January 1977 agreement was reached in principle to form a new company, Vickers-Lenning (Pty) Limited, to build and operate a foundry at Isithebe in the Bantu Homelands to produce high quality manganese steel castings: this was the project developed under the aegis of Vickers Australia, to which reference has already been made, with a Vickers majority holding of 55 per cent. In February 1977 the acquisition was announced of 88 per cent of the issued share capital of the Ernest Lowe Group of companies for a consideration of some £640,000. In addition to selling sophisticated components of hydraulic and pneumatic machinery on licence, the Lowe Group itself manufactured at Johannesburg hydraulic cylinders and other less sophisticated components. The fourth development was agreement to set up a company to manufacture Howson-Algraphy printing plates in South Africa at Pietermaritzburg, with Howson-Algraphy as majority holders and Huletts Aluminium Limited and Coates Brothers (South Africa) Limited as the other partners.

These developments meant an appreciably stronger Vickers presence in South Africa, though not approaching that established in Australia and Canada. Against the political background any further investment in South Africa had clearly to be approached with care.

In his speech at the annual general meeting in June 1976 Lord Robens included a passage on both this topic and on overseas investment generally. To criticise investment overseas was, he said, 'insularity at its most absurd'. The days were long gone when a national economy could stand on its own.

'What is required is a sensible mix of home and overseas investment. At the least, it is a form of insurance against the vagaries and recessions that occur in particular markets and particular currencies at particular times. It is also an important factor in securing a company's position in key overseas markets: and in adding to a company's overall strength and stability it can hardly fail to enhance the company's domestic performance and its ability to provide jobs at home. Incidentally, is it such a bad thing that a British company should also create jobs overseas? I was recently in South Africa. We only operate on a small scale in that country, but at least a good many black South Africans have jobs as a result—jobs available at the rate for the job regardless of colour.'

The golden rule, he concluded, whether at home or overseas, was to invest only in those projects and activities that were already earning well or had the prospect of earning well.

7

Exit Shipbuilders and BAC

To fight or not to fight, and if to fight, for total victory or for limited objectives? These were critical questions for the shipbuilding and aircraft companies faced with nationalisation when Labour won the General Election in February 1974.

Though the Queen's Speech in March contained no proposals to legislate, the intention to nationalise was declared by Mr Benn, the new Secretary of State for Industry. At the end of July he issued a 'discussion paper' on public ownership of 'shipbuilding and associated industries', and followed it with a list of companies to be taken over, including Vickers shipbuilding activities. The topic was therefore very much alive through the summer months, and the companies and their associations had to take positions. At this stage Labour was a minority government, and to try to rectify this situation it called a further election in the autumn, with nationalisation of the shipbuilding and aircraft industries included in its manifesto. The ploy succeeded but only just. Labour now had an overall majority of three. This time the Queen's Speech announced the Government's intention to introduce nationalisation legislation, and in the debate on the address an amendment by the Conservatives rejecting these 'disastrous proposals' was defeated by a comfortable majority. The course was therefore set and in January 1975 Mr Benn issued a 'consultative document' on proposals to take the aircraft industry into public ownership. In March he made a written statement setting out the Government's proposals for both industries and the Bill itself was published on 30th April, including a highly unsatisfactory formula for payment of compensation.

For the companies it was a matter of deciding whether they should seek to generate opposition to the proposals in general or to campaign against the compensation clauses. The fight in principle was already over, it could be argued. Labour's manifesto had included these proposals and the people had rendered their verdict. 'Much too faint-hearted', others might retort. 'The election was mainly on broader issues and certainly didn't mean that the people wanted everything in the manifesto. On the contrary, any public opinion poll on further nationalisation would show most people against it. And in any case the Government's majority is so small it could be toppled given a big enough effort.'

At the end of the day, however, it was simply a matter of whether the Government could get legislation through Parliament. The battle would be decided on the floor of the House. Facts, figures and arguments could be fed into the debate, and attempts made to persuade public opinion to rise in wrath, but it had to be assumed that on this issue Labour in Parliament would stand solidly together. In that case could the parties in opposition be brought to regard nationalisation of aircraft, shipbuilding and shiprepairing as the crucial issue on which they would unite? It was highly improbable given the heterogeneous views they represented and ends they sought. Nor was it likely that on this issue public opinion could be deeply stirred. The shipbuilding industry was seen as ailing and the aircraft industry as resting heavily on Government orders for military aircraft, with Concorde a controversial and extremely expensive civil project financed by the British and French Governments.

In these circumstances a close identity of view and united campaigning by the threatened companies, spread over two separate industries, could not be expected.

Some shipbuilding companies, indeed, may have felt that their survival depended on the injection of public funds, as on the Clyde, and did not really expect it to happen without State ownership. Differences in attitude also existed between the small group of warship builders and those engaged in commercial shipbuilding. The warship builders made gloomy prognostications about the effect of nationalisation on orders from foreign navies, and argued that if Government really had to take a greater degree of control it should be done by buying an equity stake in each company, in much the same way as it had done in BP. A further division in view also existed between the shipbuilders and the shiprepairers, and among the latter the Bristol Channel Shiprepairing Company, led by Mr Christopher Bailey, launched a vigorous public campaign against the inclusion of shiprepairing companies in the Bill. In all these circumstances it was perhaps not greatly surprising that the voice of the Shipbuilders and Repairers National Association seemed at times to be muted.

The Society of British Aerospace Companies, on the other hand, loudly declared its total opposition to State ownership, suggesting not least that once the main companies had been taken over it would not be long before the electronic and other equipment suppliers were dragged along the same path.

Within this ferment Vickers had a special position in that it straddled both industries and was therefore threatened by a double blow. In shipbuilding it differed from most other companies in having a lively, successful and profitable business, making no demands for public aid. Nothing could be gained in the national interest by bringing it under State control and subjecting it, in the words of Lord Robens, to 'political direction and bureaucratic control'.

Vickers had also to have regard to the interests of its wide range of businesses outside those to be nationalised. In the further expansion of these businesses much would depend on the amount received as compensation for the loss of shipbuilding and the half-share in BAC. It was therefore of great practical consequence that compensation should be 'fair'.

The trouble was that the Government's idea of fairness, as revealed in the Bill published at the end of April 1975, was poles apart from fairness as Vickers and the other threatened companies saw it. The compensation section of the Bill provided that for unquoted companies — the category covering both Vickers shipbuilding business and BAC — the amount of compensation should be calculated on what the market price would have been on given dates in the past. This was a formula with precedents but everything hinged on the choice of dates. The Bill defined the 'relevant days' as all Wednesdays during the six months beginning on 1st September 1973 (except that in Christmas week it should be Thursday, 27th December, since Wednesday was Boxing Day with the Stock Exchange closed). The argument for selecting this range of dates was that they represented the six months prior to the General Election in February 1974 so that no complaint could be made about the effect on prices of Labour's election to power. In fact, this was a period when prices were pulled down by the miners' overtime ban, the State of Emergency, the 3-day week and the threat to oil supplies from the Middle East.

For Vickers, moreover, it was a period before the BAC profits 'explosion' in 1973 and before a rise in profits from shipbuilding. BAC profits in 1972 were £6.5m. In the following four years, before nationalisation took effect in 1977, they were £13.7m, £24.2m, £30m and £40m. Corresponding figures for the shipbuilding group were £2.7m in 1972, £2m in 1973, £2m in 1974, £5.2m in 1975 and £4.7m in 1976.

Vickers public reaction to the compensation clauses was therefore to say forthrightly that if applied rigidly they represented nothing less than a formula for confiscation. The only true basis for a fair price was a valuation of assets and earnings, present and projected, at the time of acquisition, and not at some point in the past arbitrarily determined. The valuation could be made by experts and go to arbitration if necessary.

The Bill did, in fact, provide for arbitration, but linked only to the period of selected dates. The one possible hope lay in the statement that the arbitration tribunal 'in determining the base value that any securities would have had in those circumstances, shall have regard to all relevant factors'. Would 'all relevant factors' include developments in the businesses from the end of February 1974 to the time that nationalisation actually took place?

Vickers was thus concerned not only to demonstrate that the nationalisation of its shipbuilding interests and of BAC would have no industrial logic, but also that the compensation clauses if implemented rigidly would be manifestly unfair. It had also to combat the general public impression that 'Vickers' was to be nationalised.

Lord Robens accordingly wrote to all stockholders in September 1974, following the inclusion of Vickers in a list published by the Secretary of State for Industry of those shipbuilding activities it was intended to nationalise. He first established that capital employed in the Shipbuilding Group was in the order of only 18 per cent of the total employed by Vickers in the United Kingdom. Vickers would thus 'remain a major industrial company even if deprived of its shipbuilding interests'.

Given that nationalisation of shipbuilding would not destroy Vickers as a public company, he went on to say, Vickers could perhaps examine the nationalisation issue rather more objectively than would otherwise be the case. 'If it can be established that nationalisation of a given company would be in the national interest, by improving its efficiency and prospects, then common sense dictates that nationalisation should not be opposed, provided fair payment is made to the owners of the business.'

He might have added, though in this context did not, that his personal stance on the issue could certainly not be regarded as biased against the principle of nationalisation, since in the 1950s he had been a Labour Minister concerned in drafting nationalisation Bills and in the direction of several nationalised industries, and later had been Chairman of a nationalised industry for ten years.

His letter to the stockholders then asked if the efficiency and prospects of Vickers Shipbuilding Group would be enhanced by nationalisation. After examining the past and present record of the Group, and noting that it was currently building all the Royal Navy's nuclear submarines, as well as Type 42 destroyers and the new first-of-class command cruiser, he continued: 'It is clear, therefore that Vickers has acquired an international reputation for the high skills and efficiency of its shipbuilding, particularly in sophisticated shipbuilding and still more particularly in highly sophisticated naval shipbuilding. Of course, a company can gain a high reputation for the quality of its products without itself being commercially successful in the sense of being able to earn a big enough surplus to pay for the cost of the money it uses and to provide adequate funds for reinvestment and growth. The Vickers shipbuilding operation, however, has always been commercially successful and self-supporting. The only assistance it has received from the taxpayer has been that accorded to industries located in development areas and to the shipbuilding industry generally – aid, incidentally, falling short of similar supporting payments provided to the industry's European competitors. Moreover, Vickers receives no shipbuilding grants in relation to the ships it builds for the Royal Navy.'

He concluded that with nationalisation: 'Not only would the business have to be financed by the taxpayers, but the efficiency of its operation, the quality of its technology and its exporting potential would all be eroded.'

In a final summing up, the letter said: 'If we thought that the nationalisation of Vickers shipbuilders would bring real benefit to the nation, we would gladly say so. For sentimental reasons Vickers might be very unhappy to give up this activity, in which it has achieved such an outstanding international reputation, but I have made it clear that commercially Vickers Limited would survive without difficulty on the basis of its many other activities, some of them more rewarding in terms of return on capital than shipbuilding. We do believe, however, that it is not in the public interest, or in the interests of managerial efficiency and technological strength, that Vickers shipbuilding should be brought under State ownership. No serious attempt has been made, however, by those who advocate nationalisation of the business to state the grounds on which they do so other than in very generalised and ideological terms. The onus is still on them to make their case.'

The letter was duly reported in the press, and to reinforce it the Company took full page advertisements during the final months of 1975 to make four main points – 'We have never had to ask the British taxpayer to prop us up'; 'Since the 1890s Vickers have built famous ships of every conceivable kind and built them profitably'; 'We've always earned our reputation as aircraft builders'; and 'We're engineers to the world – and much more besides'. After appearing singly these four advertisements were brought together as a 4-page supplement in selected national newspapers, which apart from making a striking impact constituted a notable achievement in advertising technique.

On the specifically aircraft side, Vickers and GEC had jointly issued a statement in May 1975, following presentation of the Bill, attacking the proposals. BAC, they pointed out, was a successful, efficient and healthy company. 'It is well equipped and encompasses high standards of technology and training. It has an order book of £800m, of which £600m is for export ... No evidence has been offered or is available to show why and how the industry's operations will be improved by public ownership.' Dealing with Government arguments that the industry depended heavily on Government contracts, the statement pointed out that this was no less true of the aircraft industries in other countries. There was no novelty in Government being an important customer, but this did not mean that performance would be enhanced if customer took over manufacturer.

Much was also made in Government speeches and statements about the amount of launching aid provided for the industry. The situation took a different perspective, however, if one looked more closely at launching aid policy. This was described in a Command Paper as 'an interest-free financial contribution to the launching costs of a civil aircraft or aero-engine project, repayable as a levy on sales and licences to the extent that these are achieved ... The intention behind this arrangement is that production risks, as well as the risks of a cost overrun on development, should lie with the manufacturer, so that for example, if manufacturing costs exceed estimates and erode the actual margin, he should still pay the Government levy. Risk and the possibility of profit should thus remain with the manufacturer and provide a commercial incentive and a spur to exercise commercial judgments.'

On the arithmetic of government expenditure as a customer or in launching aid, the Government estimated in 1974 that since 1966 it had spent £1,450m (support for civil projects £300m, military research and development contracts £350m, and purchase of military aircraft and guided weapons £800m). It did not add, either that the purchase of goods certainly did not constitute 'aid', or that for this investment the industry had produced £3,158m of exports, as well as meeting UK needs in civil and defence contracts which would otherwise have required public spending in foreign currency. The nationalised industries, on the other hand, had received nearly £4,400m in subsidies in eleven years.

Argue as Vickers and others might, however, the battle was being fought, not in the public arena, but in Parliament. For the Government, getting the legislation was a matter of finding time in an overcrowded legislative programme and of feeding the right numbers into the voting lobbies at the right moments. Finding time was its principal headache, so much so that the Bill remained uncompleted at the end of the 1974–75 session and had to be reintroduced when the new session began. On going into committee it made slow progress and the Whips problems were compounded when, in May 1976, the Bill was challenged on the grounds of hybridity, with the Speaker ruling that a *prima facie* case existed. This would mean sending the Bill to a Select Committee of MPs with all that would involve in further delay. To deal with this situation the Government tabled a motion to suspend Standing Orders and so to sidestep the Speaker's ruling. The motion was won by one vote but there were 'furious scenes', to quote *The Times*, when it was alleged that the winning vote was that of a Government Whip who had broken a pairing arrangement. 'Furious scenes' became 'scenes of grave disorder' the next day. Labour Members sang the Red Flag. Mr Michael Heseltine seized the mace and waved it at the Government benches, blows were exchanged and the Speaker suspended the sitting for 20 minutes. The sequel was a re-run of the vote a month later. This the Government won by a majority of 14, having reached an accommodation with the Scottish Nationalists. In July the Bill, and others, were subjected to a guillotine motion and the Third Reading was gained by a majority of three. It then went to the Lords and now the question of hybridity was raised again, in the particular context of the inclusion of shiprepairing. It was still probable that, had the Government been prepared to withdraw the shiprepairing clauses, the Bill would have been obtained before the end of the extended 1975–76 session. In what seemed to be a fit of petulance more than anything else it chose not to do so, and to reintroduce the Bill as it stood in the 1976–77 session. The Lords now referred it to the Examiner of Trade Bills who ruled in February 1977 that hybridity existed. Though the Bill would have to be accepted by the end of the session the Government decided that this was too long to wait. 'Further delay,' it was said, 'would do grave damage to the industries which are prepared for their new future.' Accordingly an accommodation was reached with the Opposition. The shiprepairing section would be dropped and the Bill then rushed through Commons and Lords in time to be enacted during March.

This was a famous victory for Mr Christopher Bailey. He had thoroughly deserved the good luck of a legal complexity which left the Government, from their point of view, with a choice of evils. For the shipbuilders and the aircraft manufacturers, however, the moment of truth had arrived. Vesting date for the aircraft industry was fixed for 29th April and for shipbuilding 1st July.

Three years had thus elapsed since nationalisation became a serious threat in 1974. Though only the broadest estimates could be made of the amount of compensation likely to be forthcoming, Vickers had used this time to reorganise its ongoing businesses and to strengthen them by internal investment and by acquisition, financed in part by increased borrowing and in part from undistributed profits. Of necessity it had also undertaken a detailed examination of the structure and organisation of the shipbuilding group and of BAC in the light of the probable change of ownership. One problem was further invest-ment at Barrow. A plan had been developed for considerable expendi-ture on modernisation projects, but would this expenditure be regarded as a 'relevant factor' when negotiations began about compensation? Assurances on this point were sought and given, and the first stages of the work were accordingly put in hand. Again, mainly for historical and proximity reasons, a number of businesses had been placed under the aegis of the Shipbuilding Group which were not, in fact, ship-building activities. The galvanising unit sited at Palmers Hebburn was one obvious example, but Brown Brothers at Edinburgh and Slingsby at Kirkbymoorside also came into this category. It was decided therefore to combine these two latter businesses with Vickers Oceanics, the company formed in 1972 to operate a submersible charter service, into a new operating group called Offshore Engineering, and the change took effect in 1974. Control of the galvanising unit was later transferred to the Engineering Group. In the case of BAC, the merger arrangement in 1960 had provided for annual profits to be distributed to the owners, but this had not been done in recent years with the result that by the end of 1973 there was a substantial accumulation of undistributed profits. This amount was distributed in 1974 and Vickers received £5.5m. The Vickers policy of forming a properties company to handle property surplus to trading needs was also applied to BAC. Land and buildings (chiefly hangars) at Weybridge were identified as surplus to requirements and these were bought from BAC in 1974 for £8m, on the basis of an independent valuation, by Oyster Lane Properties Limited, a company formed jointly by Vickers and GEC.

The last launch at Barrow under Vickers' auspices was of HMS *Invincible* by Her Majesty The Queen. Fittingly enough, *Invincible* represented yet another 'first' for Vickers—the Royal Navy's first

16

'through-deck' cruiser, at 16,000 tons the biggest ship to be built for the Royal Navy for over a quarter of a century.

In saying goodbye to its shipbuilders, Vickers could at least note with satisfaction that 'Bill' Richardson remained the man in charge at Barrow. It could also note, with less satisfaction, that the Barrow shipyard and engineering works continued to be operated under the name Vickers – oddly enough if the Government really believed that a State corporation could run the business more efficiently than Vickers had done. For Vickers, this backhanded compliment could not really be welcome since its name was now attached to an activity over which it had no control. There was also the likelihood that in the public mind the Vickers name would still be associated with shipbuilding.

If any blessings were to be counted, it might at least be expected that Vickers would no longer come under the lash of the Arab Boycott Office because of the building at Barrow of three small submarines for Israel. There were even those who thought that Vickers' standing in the City would be enhanced by no longer having on its books the large and reputedly militant workforce at Barrow.

As to the sum to be paid in compensation, this was still to be negotiated when the vesting dates arrived. The resulting uncertainty was bound to be an inhibiting factor in forward planning. Vickers had nevertheless gone ahead strongly with the strengthening and expansion of its ongoing businesses, and to an extent had thus committed the compensation payments in advance. From this policy the benefits had already begun to flow in 1976 and 1977 and could be seen in rising profit figures. Clearly the gap created by the loss of shipbuilding and the BAC investment could not be bridged immediately, but the more sanguine prophets were even prepared to forecast that it could be closed within two or three years, given reasonably good trading conditions.

Epilogue

With my narrative coming to an end this book might also have ended had not Donald Tyerman read the manuscript and proposed a summing-up. In his view I should not shirk from 'hazarding some conclusions'.

The extent to which general conclusions can be drawn from the particular experience of Vickers is perhaps open to question. Unique is a word to use with care, but Vickers was probably unique among British industrial firms by virtue of not only having a predominant role in arms manufacture but also a major involvement in three basic industries inescapably on the shopping list of Socialist governments. Vickers therefore came under exceptional pressures. On the one hand its fortunes were linked with the twists and turns of national defence policy. On the other hand, once Labour returned to power, it could never look ahead without posing the questions: 'When will these businesses be taken from us, how will it be done and will fair compensation be paid?'

In these circumstances Vickers' achievement was to survive. It did much more than that, of course. Though by 1978 it had ceased to be 'blacksmith and armourer to the nation', it remained among British industrial leaders, with a powerful presence in three industries — mechanical engineering (including offshore engineering), office equipment and printing plates, operating in each on an international scale, with its management in high repute and its sights set on a new phase of expansion. It had also to be counted to Vickers' credit that when its interests in steel, shipbuilding and aircraft came to be nationalised all three were highly desirable assets, combining technical competence with the ability to make profits.

The story might have developed less favourably but for the time scale on which it was enacted. In the late 1950s the Company could already see that it must change direction, that it must diversify into activities with a much stronger commercial base. The process of change proved more difficult and took longer than had been hoped. So far from playing a diminishing role in weapons manufacture, Vickers found itself the sole builder of British nuclear submarines, now the capital ships of the Royal Navy, and also, through BAC, committed to the major military aircraft projects. Even in battle tank production the massive orders for Chieftain from Iran meant that Vickers' capacity remained of cardinal national importance. There was no reason, however, why this continuing load of armament work should inhibit efforts to build up activity in commercial products, except to the extent that existing capacity was still earmarked for armaments. Here the main brake at first came from lack of cash, either for re-investment or for acquisitions, because of the demands of the Vanguard and VC10 programmes and English Steel's Tinsley Park project.

Nevertheless, the policy of diversification was pursued year by year. It involved trial and error, and there were losers as well as winners. The tractor was a loser primarily because Vickers had to learn the hard way that the approach and methods used in designing and manufacturing armoured vehicles were not appropriate to a deceptively similar commercial product. The move into chemical engineering failed primarily because the classic mistake had been made of acquiring an activity divorced from the knowledge and experience of the acquirer. The disappointment in medical engineering arose primarily because the market potential overseas was misjudged and because perfecting the highly sophisticated Multichannel 300 and its derivatives took too long (though some marks are due for a brave try and for not accepting tamely the precept that other people should be left to do the pioneering). Printing machinery also proved disappointing, mainly because in a period of recession competition became intense and Vickers took very large contracts for newspaper presses on over-strict terms and conditions.

In the face of these disappointments the biggest efforts in developing new activities became concentrated on office equipment (beginning with the acquisition of Roneo), on printing plates (initially through the almost incidental acquisition of the Howson business), on offshore engineering (in particular the operation of a fleet of submersibles for work on the sea bed), and on various businesses within the Engineering Group, notably bottling machinery, bearings, machine tools and special projects. The engineering activities in Australia and Canada were also encouraged to expand.

As a corollary to the building up of new activities Vickers had to face the disagreeable task of closing down or selling businesses which for a variety of reasons, chiefly because they had outlived their markets, could no longer operate profitably in their current setting. Given the

Company's traditional paternalism (and reluctance to admit defeat), action to deal decisively with 'losers' was frequently deferred, and this gave rise to critical comment in the City. In some instances legislation on redundancy payments eventually eased the problems of closure and retrenchment, though this was not a factor in the biggest 'withdrawal' decision of all—the decision to retire from chemical engineering by sale of the Vickers-Zimmer business in Germany. The decision was almost the first taken by Peter Matthews when he came to Vickers as chief executive during the second half of 1970. It called for both courage and careful judgment for at the time the Company's finances were finely balanced and substantial losses were inevitable. In fact, the sale removed a dangerous burden, and sighs of relief echoed round the Group.

In addition to its own problems, the Vickers management shared with other industrial companies the many trials and tribulations that accumulated during the late 1960s and early 1970s. Rampant inflation, coupled with price control, obviously became a predominant anxiety. Increasing political 'interference' marked the whole period. In their efforts to control the economy Governments sought increasingly to give direction to industry, whether by outright nationalisation, by financial sticks and carrots, through parastatal organisations or by rules, regulations and requirements covering the whole range of industrial activity. It was also a period in which more and more power moved into the hands of trade unions, whose particular interests, legitimate enough, did not necessarily coincide with the broad well-being of firms and industries.

For managements the effect of these developments was to multiply paperwork, committee work and overheads, and to divert time and energy from the basic task of designing, making and selling competitively, a task sufficient in itself. Nor was management morale promoted by denigration and by the continuing erosion of incentives.

For a time in the 1960s it was fashionable to believe that detailed planning, backed by computers, could establish an industrial elysium if only stick-in-the-mud managements could be persuaded to bestir themselves. Planning in industry can never be as easy as that. There are too many imponderables, too many extraneous factors which cannot be controlled, too many changes of circumstance to achieve hard and fast certainties in plans reaching beyond a year or two. Starting with the simple objective of maintaining a viable business, there must obviously be strategic guidelines, backed by yearly planning as thorough and painstaking as it can possibly be, but wariness is still required and options kept open to seize unexpected opportunities and to cope with unexpected hazards.

In Vickers a management plan system was developed early in the 1960s with the aid of McKinsey. In theory it was difficult to fault. In practice results at first almost always fell short of projections. The trouble lay partly in accepting plans from unit managements which were not realistic, and then in failure to monitor performance adequately. Over-optimistic planning reflected in some degree at least a lack of mutual confidence and understanding between authority at the centre and authority in the operating businesses. Management at Vickers House needed to compel respect for its knowledge, judgment and expertise and also to demonstrate that sanctions existed and would be used if necessary. There was also the intangible factor of good understanding between individuals. As to monitoring, the remedy lay in monthly instead of quarterly reporting, and in centralising banking arrangements to make possible daily monitoring in Vickers House of cash flow throughout the Group. Once these weaknesses had been corrected the management plan system worked well, particularly in relation to the year immediately ahead.

Financial 'whizz kids' were also a feature of the period. Mainly they operated on the basis of making acquisitions with borrowed money and then recouping by asset-stripping. In some instances the process probably served a useful purpose, and certainly it prompted managements throughout industry to ensure that no assets were left lying idle. Vickers' own experience in 1973 and 1974 was unsettling while it lasted, but whether break-up was ever intended remains a matter for conjecture and in the end no harm was done. It was not an episode to the liking of institutional investors and they played a constructive role in stabilising the position.

Earlier a group of institutional stockholders had themselves intervened to bring about changes at the summit of Vickers. Possibly Vickers was unlucky in being selected as the guinea pig at a time when the institutions were under pressure to demonstrate investor power. It was an intervention generally regarded as clumsy, but by good fortune the hour found the man and the institutions could eventually retort: 'Well, it worked, didn't it?' And so it did—over a period—on the evidence of most of the 'before and after' statistics, including the market value of holdings. Ironically, however, the amount paid out in dividends during the five years 1972 to 1976 exceeded by very little the amount paid during the five years 1965 to 1969 (and in real terms was substantially less), principally because statutory controls came into force at a time when the Vickers dividend stood at a low figure.

In 1976, the last year before the nationalisation of aircraft and shipbuilding, profit totalled £38.3m. It was achieved from sales of over £420m and after paying over £160m in salaries and wages. Taxation took half the profit. It was also reported that during the years 1972–75 £55m had been re-invested in the existing businesses. These figures demonstrate both the achievements of the 'new' regime and the extent to which a successful private enterprise company contributes to the economic life of the nation. In one respect the regime had its 'little bit of bloomin' luck' – the surge in BAC profits which began in 1973 – but there can be no doubt about the sureness of touch with which all the prime responsibilities of leadership were exercised. These I define on page 198 as: 'Taking out losers, building up winners, getting the right people into the right jobs, and insistence on high management performance, coupled with strict financial controls.' Skill in implementing decisions was stressed by Peter Matthews as a vital management function. 'This form of management,' he said, 'is about making the best of people, about selling, about fixing prices and conditions of sale, about timekeeping, about yields, speeds, outputs and avoidance of waste, about the control of overheads, the control of stock and work-in-progress, about collecting debts, about reducing overdrafts and, above all, about influencing and controlling people.' It involved, in his view, the regular presence of managers in the places under their control: 'absentee' or remote managers rarely succeeded.

If the application of these principles helped to explain the successes of the 1970s, it is none the less important to recognise that the 'old' regime had also played its part. The change of direction imparted during the 1960s Peter Matthews described as 'decisive and bold and almost certainly right'. He also recognised that he had inherited a sound organisational structure.

For Vickers to have emerged from so many traumatic experiences still strong and with high morale is a tribute to its leadership, to its resilience and to its determination and ability to adapt. For those who believe that the well-being of society depends ultimately on the ability of a sufficient number of organisations and people to stand on their own feet the Vickers achievement deserves respect and gives encouragement.

Appendix A

VICKERS GROUP: FINANCIAL SUMMARY 1956–76

	Sales £'m	Consolidated Profit before Taxation £'m	Share of Profits of Assoc. Cos. £'m	Profit before Taxation £'m	Assets Employed £'m	Stockholders' Interest £'m	Number of Employees (5) £'m
1976	424.2	17.4	20.9	38.3	262.5	163.5	41,778
1975	324.9	18.7	15.6	34.3	239.9	154.1	41,084
1974	264.6	(1) 13.5	12.4	25.9	198.5	143.2	40,543
1973	212.7	11.4	6.8	18.2	(2) 166.3	125.0	38,199
1972	173.6	6.9	2.7	9.6	116.3	(2) 80.9	38,727
1971	181.0	4.3	1.9	6.2	111.5	76.4	39,761
1970	173.3	2.3	1.9	4.2	108.4	83.7	44,798
1969	(3) 161.4	7.6	2.3	9.9	107.1	83.2	43,348
1968	(3) 154.3	7.0	—	7.0	113.2	90.3	40,678
1967	(3) 171.7	4.7	—	4.7	118.9	90.1	45,867
(4) 1966	(3) 157.1	3.9	—	3.9	115.1	90.4	51,401
1965	(3) 185.5	6.8	—	6.8	163.1	102.3	61,214
1964	(3) 129.5	5.7	—	5.7	160.9	102.8	56,078
1963	(8) 140.0	8.5	—	8.5	166.9	109.1	56,006
1962	(8) 155.0	9.6	—	9.6	160.4	107.5	57,617
1961	(8) 169.0	9.4	—	9.4	146.7	105.4	58,531
1960	(8) 159.0	11.6	—	11.6	142.8	100.7	58,541
1959	(8) 156.0	9.1	—	9.1	138.2	99.1	(6) 70,800
1958	(8) 177.0	12.7	—	12.7	137.2	98.0	(6) 75,800
1957	(8) 203.0	13.5	—	13.5	118.9	84.0	(6)(7) 89,850
1956	(8) 163.0	12.7	—	12.7	103.8	73.4	(6)(7) 90,930

Notes

(1) After £2m provision for doubtful debts.
(2) Reflects revaluation of properties.
(3) Excluding sales of VC10 aircraft.
(4) English Steel Corporation Limited not consolidated.
(5) Excluding those employed in associated companies.
(6) Including Weybridge employees.
(7) Including Power-Samas employees.
(8) Sales in the period 1956 to 1963 were quoted in the accounts to the nearest £'m.

Appendix B

DIRECTORS OF VICKERS LIMITED
1956–77

	1956	1957	1958	1959	1960	1961	1962

The Viscount Knollys
Major-General Sir Charles Dunphie
The Lord Bicester
A. O. Bluth
Sir George Edwards
Colonel A. T. Maxwell
Sir Thomas Merton
P. H. Muirhead
Sir Frederick Pickworth
D. L. Pollock
E. J. Waddington
Lt General the Lord Weeks
Sir James Reid Young
Sir Eric Faulkner
G. H. Houlden
A. H. Hird
Sir Sam H. Brown
Sir Leslie Rowan
R. P. H. Yapp
W. D. Opher
Dr W. D. Pugh
Sir Peter Runge
J. H. Robbie
Sir Leonard Redshaw
A. M. Simmers
A. P. Wickens
Colonel H. S. J. Jelf
R. Wonfor
N. C. Macdiarmid
J. Martin Ritchie
Sir Peter Matthews
Admiral of the Fleet Sir Michael le Fanu
D. J. Groom
The Lord Robens
J. R. Hendin
W. C. P. McKie
C. W. Foreman
J. E. Harrington
F. R. P. Vinter
P. D. Scott Maxwell
D. A. S. Plastow
K. W. Ketteringham
R. O. Taylor

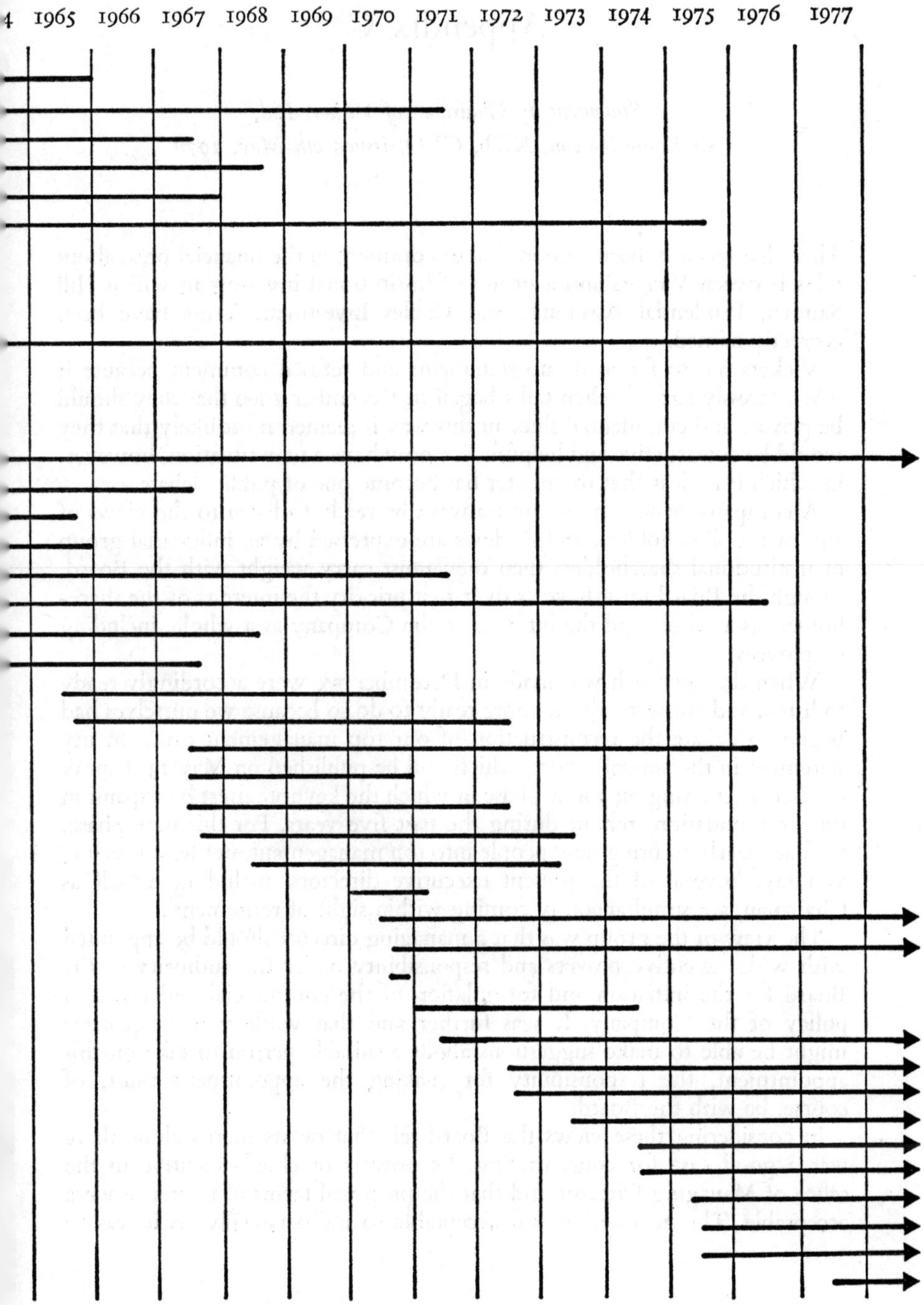
4
1965
1966
1967
1968
1969
1970
1971
1972
1973
1974
1975
1976
1977

Appendix C

Statement by Chairman of Vickers Ltd,
Sir Leslie Rowan, KCB, CVO, issued 6th May, 1970

There has recently been a good deal of comment in the financial press about
talks between Vickers and a group of institutional investors in which Hill
Samuel, Prudential Assurance and Cables Investment Trust have been
correctly named.

Vickers has so far made no statements and refused comment because it
was expressly agreed when talks began in December 1969 that they should
be private and confidential since in this way it seemed most likely that they
would be constructive and helpful. We now have a new situation, however,
in which it is clear that the matter has become one of public debate.

A company management must always be ready to listen to the views of
any of the shareholders, and if views are expressed by an influential group
of institutional shareholders then they must carry weight with the Board,
though the Board must have as their first priority the interests of the shareholders as a whole and the interests of the Company as a whole, including
employees.

When the approach was made in December we were accordingly ready
to listen, and we were all the more ready to do so because we ourselves had
begun to debate the reconstruction of our top management team. In my
statement in the annual report, which will be published on May 13, I speak
of Vickers entering on a new phase in which the keynote must be expansion
on the foundations rebuilt during the past five years. For this new phase,
we had clearly to bring new people into top management, not least because,
as I say, 'several of the present executive directors, including myself as
Chairman, are simultaneously coming within sight of retirement'.

The view of the group was that a managing director should be appointed
with wide executive powers and responsibility under the authority of the
Board for the initiation and formulation of the commercial and financial
policy of the Company. It was further said that while various quarters
might be able to make suggestions about a suitable person to take on this
appointment, the responsibility for making the appointment must, of
course, be with the Board.

In considering these views the Board felt that on its merits alone there
was a good case for concentrating the powers of chief executive in the
office of Managing Director and that the proposed terms of reference were
acceptable. The proposition was acceptable to me personally, as it was to

Mr Macdiarmid who joined us as Deputy Chairman at the beginning of this year. It is the Board's intention that Mr Macdiarmid should succeed me as Chairman at the due time.

On the question of appointing a Managing Director with enhanced powers, the Board was aided in its deliberations by the fact that Mr Yapp, who is 61, had already indicated that he would willingly stand down as Managing Director if the Board felt this would facilitate construction of the new management team and would serve the interests of the Company. No one who has had the privilege of working with Mr Yapp would be surprised by this characteristically unselfish and constructive attitude.

We accordingly began to make enquiries to identify suitable people. These enquiries ranged widely, and a list of names was compiled from which the Board finally selected Mr Peter A. Matthews, a Member of the British Steel Corporation and Managing Director, Operations and Supplies. His experience is admirably suited to the need and we have received from many sources high tribute to his character and capabilities.

Mr Matthews, who is 47, will accordingly assume the office of Managing Director, with the general terms of reference set out above, as soon as his present commitments have been cleared, which is expected to be by the beginning of September. Mr Yapp will then vacate the post of Managing Director. The Board are most anxious to retain the benefit of Mr Yapp's experience and he will remain a member of the Board as a non-executive director. He will also continue to be a non-executive director of British Aircraft Corporation (Holdings) Limited as a nominee of Vickers.

This, then, is the background to the announcement we have made today. As the new team builds up, I am sure that we can expect to see continuing the upward trend in our results which has been apparent since the poor figures in 1966. In the past five years we have spent over £32 million on acquisitions in our selected growth areas; we have carried out drastic reorganisation, involving extensive closures and disposals; we have taken on a new shape with three new groups—printing machinery and supplies, office equipment, and chemical engineering—added to shipbuilding and engineering; and our performance in terms of both profit and return on capital employed has shown consistent improvement.

Appendix D

A Letter to Stockholders of Vickers Limited
from the Chairman, The Rt Hon Lord Robens of Woldingham.

Vickers House,
19th September 1974

Dear Stockholder,

You will no doubt be aware that the Secretary of State for Industry has published a list of those shipbuilding activities proposed for nationalisation by a Labour Government, and that Vickers shipbuilding interests are named in this list.

I thought, therefore, that you would wish to know how I and my colleagues on the Board of Vickers Limited view this situation.

Shipbuilding is, of course, only one of our manufacturing activities. In the United Kingdom we have three other operating groups – engineering, office equipment and lithographic plates – and overseas we have major companies in Canada and Australia. The capital employed in the Shipbuilding Group is in the order of 18 per cent of the total employed in the United Kingdom. Vickers Limited would therefore remain a major industrial company even if deprived of its shipbuilding interests, while the payment received in compensation would be available to invest in our other areas of interest.

To that extent, nationalisation of shipbuilding would not be a mortal blow to Vickers as a public company. This perhaps makes it possible for us to examine the nationalisation issue rather more objectively than would otherwise be the case. If it can be established that nationalisation of a given company would be in the national interest, by improving its efficiency and prospects, then common sense dictates that nationalisation should not be opposed, provided fair payment is made to the owners for the business acquired.

Let us then look at the facts about Vickers shipbuilding. The Barrow Shipbuilding Works began operations in 1870 and has therefore been in the business for over 100 years. Though we have built ships of all kinds – the *Oriana*, for example, the largest liner built in England – naval shipbuilding has always been at the centre of our activities and we are the largest warship builders operating in the UK. At present we are the sole builders of Britain's nuclear submarines: in addition we are building Type 42 destroyers for the Royal Navy and also its new command cruiser. It is very much more than being builders for the Royal Navy, however, because from generation to generation many foreign navies have come to Vickers for their ships. Some

28 per cent of the orders at present in hand are for export and in 1973 Vickers shipbuilders won the Queen's Award for their achievements as exporters.

It is clear, therefore, that Vickers has acquired an international reputation for the high skills and efficiency of its shipbuilding, particularly in sophisticated shipbuilding and still more particularly in highly sophisticated naval shipbuilding. Of course, a company can gain a high reputation for the quality of its products without itself being commercially successful in the sense of being able to earn a big enough surplus to pay for the cost of the money it uses and to provide adequate funds for reinvestment and growth. The Vickers shipbuilding operation, however, has always been commercially successful and self-supporting. The only assistance it has received from the taxpayer has been that accorded to industries located in development areas and to the shipbuilding industry generally – aid, incidentally, falling short of similar supporting payments provided to the industry's European competitors. Moreover, Vickers receives no shipbuilding grants in relation to the ships it builds for the Royal Navy.

Against this background we have to ask ourselves whether nationalisation would result in a more economic operation, whether it would produce improved management efficiency, whether it would generate a higher rate of technological progress and whether it would give us a stronger position in foreign markets.

In these respects I suggest that the reverse would be more likely to be true. Not only would the business have to be financed by the taxpayers, but the efficiency of its operation, the quality of its technology and its exporting potential would all be eroded.

So far as building for the Royal Navy is concerned, the Government would be trading with itself. This could damage the economic running of the business, since there would no longer be an imperative need to arrive at realistic contracts and to keep within their financial disciplines. This damage would be aggravated by the removal of control from the men on the spot to a distant policy-making body, motivated to some degree at least by extraneous considerations. It would also be likely to slow down the rate of technological advance since this depends so much on the creatively abrasive dialogue between the customer, with his desired requirements, and the builder, who has to be ingenious and practical in meeting these requirements or in offering alternatives.

Then there is the question of exports. I have said that some 28 per cent of our shipbuilding order book is represented by work for foreign navies. Indeed, I understand that 60 per cent of all orders for naval vessels placed overseas by foreign governments come to Britain, so that this is a very important element in the balance of trade. It is questionable, to say the least, whether foreign governments would be as willing to place orders with British State-owned yards as they are with British companies on a commercial basis with commercial secrecy built into contracts.

In short, it can only be expected that nationalisation would reduce the commercial and technological efficiency and prospects of the shipbuilding business built up by Vickers over several generations, and so reduce the value of the national asset that Vickers shipbuilding represents. In the longer run a decline in the business could lead to reduced employment opportunities at Barrow-in-Furness, a town heavily dependent on shipbuilding for its prosperity and well-being. Vickers at Barrow today employs some 14,000 people, an increase of 50 per cent in shipbuilding in the last ten years. The maintenance of a lively and successful business, with the ability to win foreign orders, is vital to Barrow, quite apart from the considerations of national interest that I have outlined.

If we thought that the nationalisation of Vickers shipbuilders would bring real benefit to the nation, we would gladly say so. For sentimental reasons Vickers might be very unhappy to give up this activity, in which it has achieved such an outstanding international reputation, but I have made it clear that commercially Vickers Limited would survive without difficulty on the basis of its many other activities, some of them more rewarding in terms of return on capital than shipbuilding. We do believe, however, that it is not in the public interest, or in the interests of managerial efficiency and technological strength, that Vickers shipbuilding should be brought under State ownership. I have explained why we believe this. No serious attempt has been made, however, by those who advocate nationalisation of the business to state the grounds on which they do so other than in very generalised and ideological terms. The onus is still on them to make their case.

Yours sincerely,

ROBENS

Appendix E

Ships and Submarines delivered and launched at the Vickers Shipyards
in the United Kingdom 1956–77

WARSHIPS

Aircraft carrier: HMS *Hermes*

Frigates: HMS *Scarborough*
HMS *Eastbourne*
HMS *Mohawk*
HMS *Penelope*
HMS *Minerva*

Beas (India)
Betwa (India)

Rostam (Iran)
Zaal (Iran)

Destroyers: HMS *Glamorgan*
(Guided missiles destroyer)
HMS *Sheffield*
(First of class: Type 42 Seadart
destroyer)

Aragua (Venezuela)
Almirante Williams (Chile)
Almirante Riveros (Chile)
Hercules (Type 42: Argentine)

Cruiser: HMS *Invincible*
(First of class anti-submarine
cruiser, launched 1976)

Corvette: *Keta* (Ghana)

SUBMARINES

HTP: HMS *Explorer*
HMS *Excalibur*

Porpoise Class: HMS *Porpoise*

HMS *Rorqual*
HMS *Narwhal*

Oberon Class: HMS *Orpheus*
HMS *Olympus*
HMS *Osiris*
Humaita (Brazil)
Tonelero (Brazil)
Riachuelo (Brazil)

Nuclear: HMS *Dreadnought*
(First British)
HMS *Valiant*
(First all–British)
HMS *Warspite*
HMS *Churchill*
HMS *Courageous*
HMS *Swiftsure*
HMS *Sovereign*
HMS *Superb*
HMS *Sceptre*
(launched 1976)
HMS *Spartan*

Polaris: HMS *Resolution*
(First British ballistic-armed)
HMS *Repulse*

Coastal: *Gal* (Israel)
Tanin (Israel)
Rahav (Israel)

LINERS

For Canadian Pacific Steamships
Limited:
Empress of England
Empress of Canada

For Orient Steam Navigation
Company:
Oriana
(42,000 tons: largest liner
built in England)

For Shaw Savill Company:
Northern Star

For Black Sea Shipping Company:
Odessa

CARGO LINERS

For Ellerman Lines Limited:
City of Ripon
City of Auckland
City of Eastbourne
City of Glasgow

For Furness Withy Company:
Pacific Envoy
Pacific Stronghold

For Alfred Holt & Company:
Anterior
Achilles
Ajax
Priam
Peisander
Prometheus
Protesilaus
Radnorshire

For Shaw Savill Company:
Illyria

For Canadian Pacific Steamships
Limited:
Beaveroak

TANKERS

For Niarchos Group:
Spyros Niarchos
Evgenia Niarchos

For Shell Tankers Limited:
Hinea
Hinnites
Serenia
(1961, 67,000 tons, largest
tanker then built in Britain)

For BP Tankers Company
Limited:
British Glory
British Faith
British Ambassador
British Prestige
British Grenadier
British Admiral
(1965, first 100,000 ton tanker
built in Europe)

For Esso Petroleum Company
Limited:
Esso Durham
Esso Portsmouth
Esso Edinburgh
Esso Newcastle

For Eagle Tanker Company:
San Gregorio

For Alvion Steamship Company:
Alvenus

For Jorgen P. Jensen:
Canto

For Orient Line:
Garonne

For Charter Shipping Company
Limited:
Malwa

For Regent Petroleum Tankship
Company Limited:
Regent Pembroke

OTHER VESSELS

Memnon, cargo ship for Alfred Holt & Company
Laurentic, refrigerated cargo ship for Shaw Savill & Albion Company
  Limited
Aranui, car–rail–passenger ferry for Government of New Zealand
Methane Princess, liquefied methane gas carrier for Conch Methane Tankers
  Limited
Albright Pioneer ⎫
Albright Explorer ⎬ Phosphorus carriers for Albright & Wilson Limited

Appendix F

Baron	1956	LT. GENERAL SIR RONALD WEEKS, KCB, CBE, DSO, MC, TD, lately Chairman, Vickers Limited.
Knights	1957	G. R. EDWARDS, CBE, BSC, Managing Director, Aircraft Division, Vickers-Armstrongs Limited.
		F. PICKWORTH, Chairman, English Steel Corporation Limited.
	1959	MAJOR-GENERAL C. A. L. DUNPHIE, CB, CBE, DSO, Managing Director, Vickers Limited.
	1972	LEONARD REDSHAW, Assistant Managing Director, Vickers Limited, and Chairman, Vickers Limited Shipbuilding Group.
	1975	P. A. MATTHEWS, Managing Director, Vickers Limited.
CBE	1957	E. L. CHAMPNESS, Managing Director, Palmers Hebburn Company Limited.
	1960	CAPTAIN G. I. D. HUTCHESON, Managing Director, Cockatoo Docks & Engineering Company Pty Limited.
	1963	W. D. OPHER, Chairman and Managing Director, Vickers-Armstrongs (Engineers) Limited.
	1965	W. D. PUGH, Managing Director, English Steel Corporation Limited.
	1977	J. R. HENDIN, Assistant Managing Director, Vickers Limited, and Chairman, Vickers Limited Engineering Group.
OBE	1957	A. WRAGG, Manager, Shell Department and Chief Metallurgist, Vickers-Armstrongs (Engineers) Limited, Newcastle.
	1959	MAJOR P. L. TEED, Deputy Director, Department of Aeronautical Research and Development, Vickers-Armstrongs (Aircraft) Limited.
		G. R. BRYCE, Chief Test Pilot, Vickers-Armstrongs (Aircraft) Limited.
	1961	J. MELVILLE, Technical Manager, Vickers-Armstrongs (Shipbuilders) Limited, Barrow.

1963 A. L. WHITE, Director, Vickers-Armstrongs (Shipbuilders) Limited and General Manager, Palmers Hebburn Works.

1964 A. H. SULLY, Director, English Steel Corporation Limited.

1965 R. F. W. KEAY, Director of Production Engineering, Vickers-Armstrongs (Engineers) Limited.

1968 J. S. BATY, lately Chief Engineer, Internal Combustion Engine Department, Vickers Limited Shipbuilding Group.

1970 G. STANDEN, Engineering Production Manager and Local Director, Vickers Limited Shipbuilding Group, Barrow.

1971 L. G. GOOCH, Manufacturing Director, Roneo Limited, Dartford.

1972 E. A. ROBERTS, Nuclear Submarine Support Manager, Vickers Limited Shipbuilding Group, Barrow.

1973 J. A. WILSON, Chief Electrical Engineer, Vickers Limited Shipbuilding Group, Barrow.

1975 F. P. TANNER, lately Managing Director, Howson-Algraphy, Leeds.

MBE 1956 H. F. BURRELL, lately Manager, General Engineering Department, Barrow.
G. L. CARDWELL, Chief Die Designer, English Steel Forge & Engineering Corporation Limited.

1957 H. RENTON, Manager, Building Department, Vickers-Armstrongs (Engineers) Limited, Barrow.

1959 C. ROWLEY, Assistant Chief of Armament Design, Vickers-Armstrongs (Engineers) Limited.
E. BERRY, Deputy Manager, Contracts Department, Vickers-Armstrongs Limited.
F. H. SMITH, Director of Optical Design, C. Baker Instruments Limited.

1960 D. RODGER, Manager, Marine Department, Vickers-Armstrongs (Engineers) Limited, Barrow.

1961 H. E. HALL, Designer, Armament Drawing Office, Vickers-Armstrongs (Engineers) Limited, Barrow.

1962 W. J. HERITAGE, Chief Progress and Planning Officer, Forge, English Steel Corporation Limited.
A. G. SAMPSON, Ship Manager, Vickers-Armstrongs (Shipbuilders) Limited, Newcastle.

1963 F. TRAVIS, Chief Designer, Naval Armament, Vickers-Armstrongs (Engineers) Limited, Newcastle.

1964 G. STANDEN, Marine Installation and Commissioning Manager, Vickers-Armstrongs (Shipbuilders) Limited, Barrow.

1965   W. BELL, Manager, Naval Gun Mounting Department, Vickers-Armstrongs (Engineers) Limited, Newcastle.

M. C. OLDHAM, Chief Chemist and Metallurgist, Vickers-Armstrongs (Engineers) Limited, Barrow.

J. E. SWINBURN, Departmental Manager, Armament Forgings and Services Liaison Officer, English Steel Corporation Limited.

1967   E. CUSSANS, Works Manager, Vickers Instruments.

W. G. HALL, Technical Manager (Naval Armaments), Vickers Limited Engineering Group, Newcastle.

1968   G. L. CHAPMAN, Chief Designer, Armament Division, Vickers Limited Engineering Group, Barrow.

J. FORMAN, Marine Installation Department, Vickers Limited Shipbuilding Group, Barrow.

1970   D. GARDINER, Machining Manager, Armament Division, Vickers Limited Engineering Group, Newcastle.

D. CLAYTON, Engineering Manager, Vickers Limited Shipbuilding Group, Barrow.

1971   D. WALLACE, Production Manager, Armament Division, Vickers Limited Engineering Group, Newcastle.

1972   E. GILBERT, Technical Manager, Marine Engineering, Vickers Limited Shipbuilding Group, Barrow.

H. HOGGARD, Service Manager, Brown Brothers and Company Limited, Edinburgh.

H. BARKER, Manager, Fabrication Departments, Manufacturing Division, Vickers Limited Shipbuilding Group, Barrow.

R. BREWIS, Contracts Manager, Armament Division, Vickers Limited Engineering Group, Newcastle.

G. H. WILKINSON, lately Outside Erection Manager, Armament Division, Vickers Limited Shipbuilding Group, Barrow.

1973   H. SPENCE, Technical Manager, Armament Division, Vickers Limited Engineering Group, Newcastle.

1975   E. CLEMENTS, Chief Estimator, Barrow Shipbuilding Works, Vickers Limited.

H. O. CLEMENTS, Manager, Proving and Testing, Armament Division, Vickers Limited Engineering Group, Newcastle.

R. J. HICKS, General Manager and Local Director, Compact Orbital Gear Works, Rhayader.

A. H. JOLLY, Chief Test Engineer, Vickers Limited, Barrow Engineering Works.

W. E. STABLER, lately Leading Checker, Vickers Limited Engineering Group Newcastle.

1976   B. SPARK, Personal Assistant, Vickers Limited, Newcastle.

1977 Miss Elizabeth Head, Senior Clerk, Wages Department, Vickers Limited Engineering Group, Newcastle.

BEM 1956 G. G. Purvis, Inspection Foreman, Elswick Works, Newcastle.

G. W. Richardson, Toolroom Inspector, Weymouth Works.

1957 V. H. Crozier, Honorary Collector for National Savings, Barrow.

E. Fleming, Head Foreman, Barrow.

R. Smithers, Senior Foreman, Cooke, Troughton & Simms Limited.

J. H. Laing, Skilled Machinist, English Steel Corporation Limited.

1958 W. H. Butler, Apprentice Instructor, Supermarine Works, Vickers-Armstrongs (Aircraft) Limited, Southampton.

W. A. Wright, Manager (No. 7 Works), Punched Card Factory, Powers-Samas Accounting Machines Limited, Croydon.

1959 S. Couch, Head Foreman Coppersmith, Vickers-Armstrongs (Shipbuilders) Limited, Newcastle.

1960 C. E. Wright, Assistant Manager, Galvanising Department, Vickers-Armstrongs (Shipbuilders) Limited, Palmers Hebburn Works.

J. G. Anderson, Assistant Foreman Electrician, Vickers-Armstrongs (Shipbuilders) Limited, Barrow.

1961 W. G. Mason, Head Foreman Welder, Vickers-Armstrongs (Engineers) Limited, Barrow.

1962 T. R. Yates, Fitter, Vickers-Armstrongs (Engineers) Limited, Newcastle.

W. H. Quayle, Caulker-Burner, Vickers-Armstrongs (Shipbuilders) Limited, Barrow.

T. Entwhistle, Chargehand Draughtsman, Vickers-Armstrongs (Shipbuilders) Limited, Barrow.

W. W. F. Fry, Planner and Estimator, Vickers-Armstrongs (Engineers) Limited, Weymouth.

A. G. Hill, Marker-Off, Vickers-Armstrongs (Engineers) Limited, Crayford.

W. Spencer, Chargehand Caulker, Vickers-Armstrongs (Shipbuilders) Limited, Barrow.

1963 J. W. Danson, Chargehand Plater, Vickers-Armstrongs (Engineers) Limited, Barrow.

H. W. Fenwick, Assistant Foreman Fitter, Vickers-Armstrongs (Shipbuilders) Limited, Barrow.

1964 A. C. Jones, Head Foreman Fitter, Vickers-Armstrongs (Engineers) Limited, Barrow.

1966   J. J. M. SHEPHERD, Foreman Coppersmith, Vickers
       Limited Engineering Group, Armament and Commercial
       Engineering Division, Newcastle.
       J. J. THOMPSON, Plater, Vickers Limited Shipbuilding
       Group, Barrow.

1967   W. DEVINE, Foreman Machinist, Vickers Limited
       Engineering Group, Newcastle.
       H. DICKINSON, Assistant Manager (Electrical) Vickers
       Limited Shipbuilding Group, Barrow.
       F. GLASS, Caulker, Vickers Limited Shipbuilding Group,
       Newcastle.
       E. SMITH, Foreman Fitter, Vickers Limited Engineering
       Group, Newcastle.
       J. T. VERNON, Chief First Aid Room Attendant, English
       Steel Corporation Limited.

1968   MISS LILIAN HORNBY, Supervisor, Mechanical Accounts
       Department, Vickers Limited Engineering Group,
       Barrow.
       T. METCALFE, Section Leader, Submarine Engineering
       Drawing Office, Vickers Limited Shipbuilding Group,
       Barrow.

1969   G. BARROW, Head Foreman Coppersmith, Vickers
       Limited Shipbuilding Group, Barrow.
       A. J. PETERS, Head Foreman, Main Assembly Shop,
       Vickers Limited Engineering Group, Crayford.

1972   H. CLARK, Foreman, Test Engineers, Vickers Limited
       Shipbuilding Group, Barrow.
       R. GIBSON, Senior Liaison Officer, Armament Division,
       Vickers Limited Engineering Group, Newcastle.

1973   T. TYRELL, Fitter, Brown Brothers and Company
       Limited, Edinburgh.

1974   H. JEVONS, Leading Launchway Shipwright, Vickers
       Limited Shipbuilding Group, Barrow.
       P. MESSENGER, Standards Inspector, Howson-Al graphy
       Leeds.

1976   A. C. CALLISTER, Painter, Vickers Limited Shipbuilding
       Group, Barrow.
       H. Y. SLATER, Foreman Fitter, Armament Division,
       Vickers Limited Engineering Group, Newcastle.
       S. CULBERT, Chargehand Fitter, Vickers Engineering
       Works, Barrow.
       F. McBRIDE, Woodcutting Machinist, Vickers Limited
       Shipbuilding Group, Barrow.

1977   J. J. McPHILLIPS, lately Foreman, Sheet Metal
       Department, Vickers Limited Shipbuilding Group,
       Barrow.

A. GRANT, Electric Welder, Defence Systems Division, Vickers Limited Engineering Group, Newcastle.

The following Honours were conferred on non-executive Directors of Vickers Limited and subsidiary companies during this period.

KBE   1956   SIR THOMAS MERTON
OBE   1968   A. O. BLUTH (also Chairman, Barber-Greene Olding and Company Limited).
Knight Bachelor
               1974   G. B. KATER, (Vickers Holdings Pty. Limited and Vickers Australia Limited).
               E. O. FAULKNER, MBE

Lieutenant-Colonel H. L. Carey, The Parachute Regiment Territorial Army, was appointed an OBE (Military Division) in 1966 (then Managing Director, Forge and Foundry Division, Vickers Limited Engineering Group, Newcastle.)

Appendix G

In an unpublished manuscript John W. Wilkinson, Chief of Armament Design at Barrow when he retired in the late 1940s, recalls the Yard as it was at the beginning of the century, shortly after Vickers had taken over.

In those days, he says, building warships was largely a matter of muscle.

'Holes for doors or ports had to be drilled by hand, with small holes round a marked-off outline. Each hole had to be perforated quite clearly without either running into its neighbour or leaving too much metal. After this had been done the hand chipper would take over and cut the plate out.

'No welding being possible, everything had to be secured by rivets or bolts. Men spent their lives drilling holes and others by continually riveting. No electric lights were strung about the dark corners of the ship's decks. Candles were the usual light and these flared and guttered over the bottles which for the most part formed the candlesticks. For greater light paraffin flares were common. The light was invariably so dim that they had to be held close to the work, and the shadows cast were the cause of accidents.'

After describing how new shops were built during the first decade of the century, with machines and cranes electrically operated, he continues: 'In the old shops there were large gas engines bolted on the walls. The shop floors were all of earth, often saturated with oil or soluble cutting mixtures which imparted a rancid odour, perhaps encouraged by the practice of mending the floor with cast iron machine cuttings. Candles were in constant use, any general lighting coming from paraffin flares or the uncertain gas flames from fish-tailed burners. In winter the old shops were dark, draughty and smoky. No practice existed of providing hot water for tea. Before the invention of thermos flasks tea was brought in already brewed and in cans, to be warmed up on one of the many fires.'

Wilkinson also had something to say about the personalities of those days.

Trevor Dawson (later knighted) had been persuaded by Albert Vickers to join Vickers in 1896 from the Royal Navy. Wilkinson recalls him as a naval gunnery specialist, and speaks of the importance and influence of a paper read by Dawson to a scientific society in 1906 on 'The Development of Ordnance'.

In his turn Dawson drew into Vickers Charles Craven, a submarine officer, the man who in due course became chairman of Vickers-Armstrongs

Limited and Vickers Limited chief executive. Wilkinson remembers Craven as 'a tall man, with a slight stoop and a practice of never seeming to hurry. These features gave him a thoughtful appearance, with an apparent capacity for bearing responsibility. He showed techniques of boldness in securing shipbuilding orders in the face of the keenest competition, and used to say "that he knew a good deal about ships, but not about shipbuilding, but he was a good carpet-bag salesman and if he could get his foot in the door he was difficult to move without an order".'

Craven's vigorous salesmanship was particularly important to Barrow in the years immediately following the First World War. 'The workshops which employed more than 4,000 men in the armament department now employed 500,' records Wilkinson. 'The whole Yard, which in normal times employed 15,000 men, and recently had more than 30,000 on its books, was down to 4,000. All production work on armaments was stopped, and instructions were received from the Ministry of Munitions to scrap and break up certain unfinished parts. The howitzer shop, with 14 bays, several hundred machines and facilities for some hundreds of men, was entirely closed down, the cranes overhead seeming by their inactivity to accentuate the silence.'

But Craven 'showed he was as good as his word'. Instead of warships Barrow built liners, even if at small profit. Foreign navies came to buy submarines. The armament drawing office was kept open, and this proved important as the tide began to turn. 'In a year or two Barrow Yard became envied by many.'

Wilkinson gives several striking pen pictures of other Vickers personalities of the day, notably of James Horne, the chief designer at Barrow.

'He was an austere man to his assistant and subordinates,' says Wilkinson, 'but in modern terms he had more than the usual share of "playboy" in his composition. The only indication of this in the office was a gardenia in his lapel and a strong scent of attar of roses which lingered in the air as he moved about the office. These occasions were rather spasmodic and were generally associated with answers required by Admiralty or London office with some urgency. The urgency of his movements was a spur to his assistants, who flew about the office as in fear of their lives, much probably to his silent amusement. He had a great reputation as a hydraulic engineer and patentee, a reputation which grew in Barrow Works to mythical form.'

Even then there was the closest liaison with the Government's ordnance experts.

'At Barrow a resident Admiralty engineer officer was in charge of day-to-day inspection of all Admiralty orders of gun mountings, while field equipments, garrison mountings and military supplies were under War Office control. Trained inspectors from the respective branches of the Services carried out detailed inspections, recordings and stamping and testing of materials.'

Though Wilkinson's manuscript makes many comments of general interest, it is essentially concerned with the technicalities of the design and manufacture of guns, together with their mountings and control systems. Clearly he was very proud of Vickers' contribution to the country's armament needs, and his own part in it was recognised by the award of an OBE.

Index

Compiled by Michael Gordon

Note. Bold type indicates biographical or historical information.